THE ROLE OF TREES IN SUSTAINABLE AGRICULTURE

FORESTRY SCIENCES

Volume 43

The titles published in this series are listed at the end of this volume.

The Role of Trees in Sustainable Agriculture

Review papers presented at the Australian Conference,
The Role of Trees in Sustainable Agriculture,
Albury, Victoria, Australia, October 1991

Edited by
ROSLYN TAMARA PRINSLEY
Manager, Research and Strategy,
Rural Industries Research and Development
Corporation, Canberra, Australia

Compiled by
JAMIE ALLNUTT

Reprinted from *Agroforestry Systems*, Volume 20, Nos. 1–2 (1992)

SPRINGER-SCIENCE+BUSINESS MEDIA, B.V

Library of Congress Cataloging-in-Publication Data

Australian Conference, the Role of Trees in Sustainable Agriculture
(1991 : Albury, N.S.W.)
The role of trees in sustainable agriculture : review papers presented at the Australian Conference, the Role of Trees in Sustainable Agriculture, Albury, Victoria, Australia, October 1991 / edited by Roslyn Tamara Prinsley ; compiled by Jamie Allnutt.
p. cm. -- (Forestry sciences ; v. 43)
ISBN 978-94-010-4809-5 ISBN 978-94-011-1832-3 (eBook)
DOI 10.1007/978-94-011-1832-3
1. Agroforestry--Australia--Congresses. 2. Tree crops--Australia--Congresses. 3. Sustainable agriculture--Australia--Congresses.
I. Prinsley, Roslyn Tamara. II. Allnutt, Jamie. III. Title.
IV. Series.
S494.5.A45A92 1991
634.9'9'0994--dc20 92-36592

ISBN 978-94-010-4809-5

Printed on acid-free paper

Originally published by Kluwer Academic Publishers in 1993
Softcover reprint of the hardcover 1st edition 1993

Table of Contents

The Conference Organising Committee is grateful to the following sponsors:

Rural Industries Research and Development Corporation
National Afforestation Program
Australian Newsprint Mills (ANM, Albury)
Greening Australia
Bureau of Rural Resources
Treecrops
Sure Gro (Cheltenham, Victoria)

Agroforestry Systems **20**: vii, 1992.

Preface

Agroforestry has the potential to become a widespread from of land use, changing and improving the sustainability and profitability of agriculture and forestry. Trees are used on farms for shelter, shade, production of timber, fodder and other products, and for conservation of biodiversity and soil. Trees can be used simultaneously for all of these benefits. Uses of trees on farms will vary depending on the particular farm's physical environment, the farmer's economic and other aims and relevant Government policies.

Thus, agroforestry, unlike many other rural industries, is a multi-faceted, multi-purpose land use. In the wet temperate zones of Australia, it is more likely to refer to commercial timber production from trees which are integrated with the farm in some way, such as timberbelts. However, in rangelands, in the semi-arid zone, it may mean the use of saltbush for fodder. In many cases, agroforestry may have no direct product such as timber associated with it, but rather the trees may be used, for example, to lower watertables in order to prevent salinisation or to act as wildlife corridors.

In many parts of Australia the benefits of trees on farms are not well understood, and agroforestry as a land use is very much in its infancy. For any land use to expand successfully, appropriate research and development is necessary.

Agroforestry research and development in Australia has been largely fragmented and many of the research results have never been published and are unknown. The purpose of this volume is to comprehensively review all of the research that has taken place in the field of agroforestry in Australia including previously unpublished results, providing readers with the latest technical and economic information about using trees in agriculture for controlling salinity and erosion, for shelter and shade, and for production of timber, fodder and minor forest products. The chapters in the book provide information concerning planted trees within all of these categories and include a special review of the management of native vegetation on farms. These papers also examine research needs where appropriate.

This book stems from the National Australian Conference on 'The Role of Trees in Sustainable Agriculture' which took place in Albury, Victoria, Australia in October 1991. Each national review paper is based upon a summary of six or more state review papers of research and development prepared specifically for the national conference. The book thus provides readers with a comprehensive overview of agroforestry research in Australia which is introduced and summarised in the first chapter.

Roslyn Tamara Prinsley

Agroforestry Systems **20**: 1–23, 1992.

Tree planting for dryland salinity control in Australia

N. J. SCHOFIELD
Western Australian Water Resources Council, PO Box 100, Leederville, Western Australia 6007

Key words: salinity, salinisation, dryland salinity, irrigation salinity, tree, reforestation, plantation, agroforestry, water, water table, groundwater, social, political, economic, financial

Abstract. Dryland salinity is emerging as a major form of land and water degradation in southern Australia, particularly in Western Australia, South Australia and Victoria, and to a lesser extent in New South Wales, Tasmania and Queensland.

Tree planting, in combination with other vegetation treatments, is regarded as a leading solution to dryland salinity. Research has now shown that planting trees can significantly lower groundwater tables, and thereby reverse the causal process of salinisation. Substantial progress has been made towards answering the basic questions of which species to grow, how to plant, where to plant, at what density and configuration to plant, and how much area to plant. The economic potential for commercial tree planting has given impetus to partial reforestation in higher rainfall (>600 mm yr^{-1}) areas. Even so, serious constraints are apparent, relating primarily to cost, uncertainty and attitude. The future should bring increasing community input and control, technical refinement and gradual adoption, provided all aspects of the issue are well researched, relevant information is well communicated and programmes are well administered. Some specific requirements for future research are identified in this review.

1. Introduction

Secondary or man-induced salinisation is of growing concern in southern Australia and in limited areas of northern Australia. This concern arises because salinity usually decreases the productive potential of the land, decreases the water resources available for consumption, and degrades the ecological values of wetlands, rivers and their associated habitats.

Secondary salinisation is a by-product of replacing native vegetation with dryland agriculture or irrigated agriculture. Observations of salinisation following clearing of native vegetation go back to the turn of the century and earlier [65]. However, it is only in relatively recent times that the community has begun to respond seriously to this slowly emerging problem. Even now there is a good deal of scepticism about whether salinisation can and need be controlled.

Considerable progress has been made toward understanding the mechanisms of salinisation and determining solutions [e.g. 28, 35, 54, 55]. This research has led to the increasing acceptance of tree planting (in combination with other vegetation measures) as a preferred option for control of dryland salinity and as having a potential role in controlling irrigation salinity [26].

However there still remains a lack of information, uncertainty and difference of opinions on basic questions such as: which species to plant, where to locate plantings, at what densities to plant, and how much area to plant. This situation has contributed to the reticence of some governments and communities to act.

Perhaps over-riding all the technical uncertainties has been the question of cost and benefit. Sustainable rehabilitation can be expensive, and the benefits are not always substantial or immediate in financial terms. A commonly expressed difficulty is the concept of replanting prime agricultural land to trees to restore downslope salinised areas, or at least to control the spread of salinisation. Unless the planted trees have high value and can be harvested, or it can be demonstrated that the land to be reforested will ultimately go saline, there is little incentive for landholders to reallocate precious, productive land.

If sound technical and economic arguments for tree planting can be made, there remains the question of how best to implement such programmes. How can government bureaucracies and local communities become motivated, who makes decisions on what, where, when and how much planting; and how can financial and human resources be increased? Once programmes are formulated, what will be the social impacts in their implementation, and how would this compare to their absence?

The foregoing raises some of the leading questions in this important issue. This paper attempts some generalised statements relating to the technical, economic, social and administrative aspects of tree planting programmes. As such, it is a subset of the broader question of how whole farm systems can be developed for sustainable land use. The States have progressed to greater and lesser degrees in these areas as can be seen from the State and general papers [8, 9, 19, 39, 56, 62] which have contributed to this review. The task here is not to identify the strengths and weaknesses of each State's programmes, but rather to draw together common experiences and views whilst utilising differing views as a means of self-criticsm.

2. Extent of secondary salinisation in Australia

2.1. *Current situation*

Areas of Australia that have been identified as hosting or being susceptible to dryland salinity are shown in Table 1 and Fig. 1. In terms of area, southern Western Australia is most affected, followed by South Australia, Victoria, New South Wales, Tasmania and Queensland. The ranking in terms of cost impact is not available.

Some degree of correlation between native vegetation clearing in Australia (Fig. 2) and dryland salinity (Fig. 1) is evident. Although such a correlation

Table 1. Areas of dryland secondary salinity in Australia.

State	Area (ha)	Source
New South Wales	12,000	[58]
Queensland	8,000	[64]
South Australia	250,000	[18]
Tasmania	11,000	[16]
Victoria	55,000*	[20]
Western Australia	440,000	[19]

* [64] quotes 100,000 ha, which includes incipient salinity.

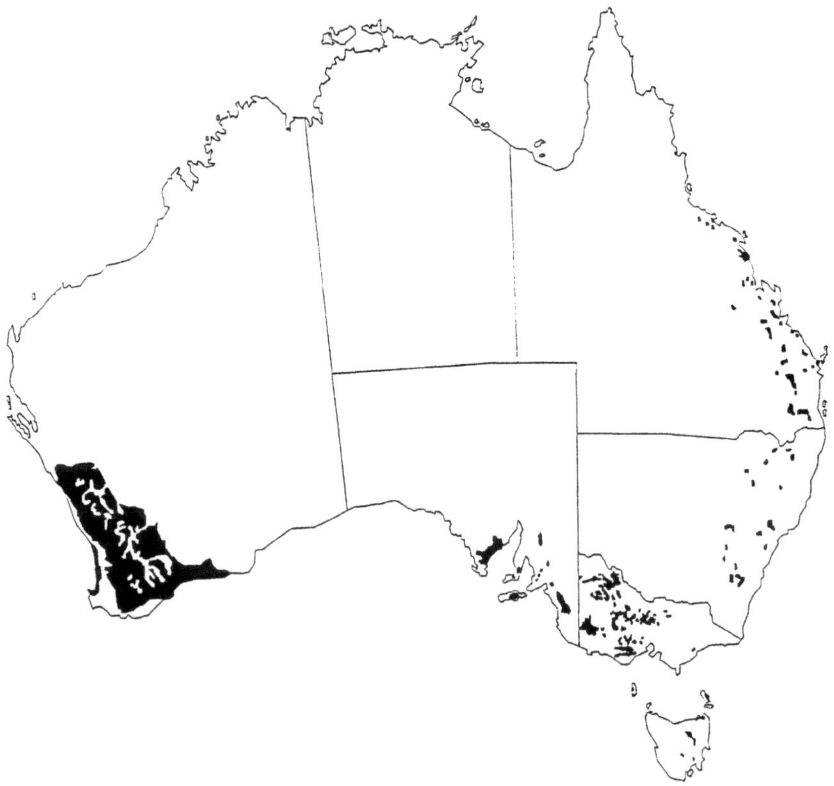

Fig. 1 Areas susceptible or containing dryland salinity in Australia (figure should be considered schematic as definitions of dryland salinity areas vary from State to State) [Sources: 8, 19, 39, 66].

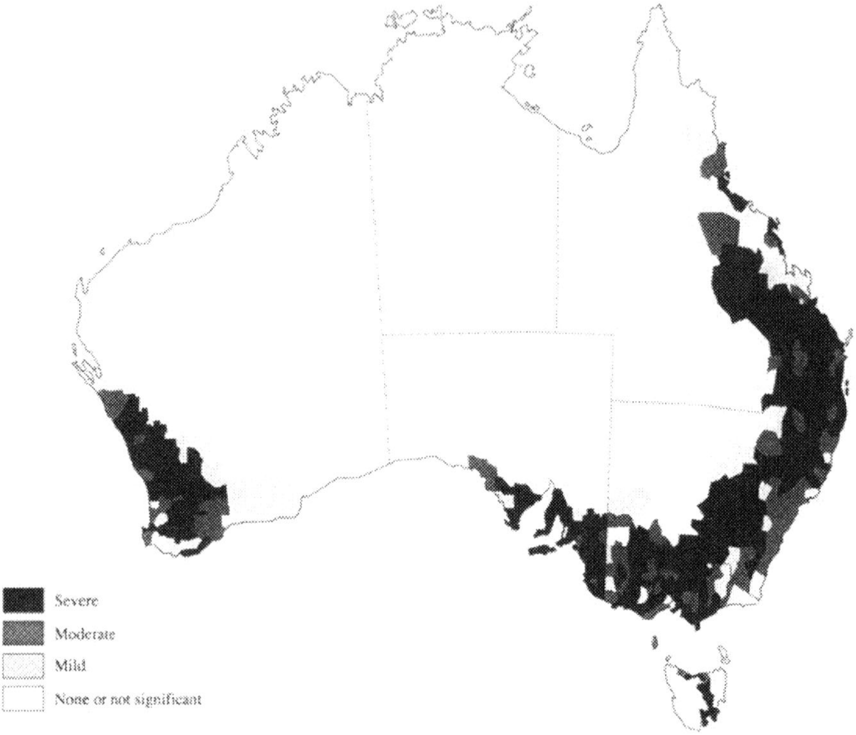

Fig. 2. Extent of vegetation clearing since European settlement of Australia (Source: John Ive, CSIRO Division of Wildlife and Ecology, pers. comm., 1991).

follows from the causal mechanisms, it is by no means the case that salinity always follows clearing.

2.2. *Future trends*

Observations from most States indicate that the area of salt-affected land is increasing. This is again most evident in south-western Australia where saltland is estimated to be increasing at 18,000 ha yr^{-1} [50] and could cover 7% of agricultural land by the year 2100 [20]. In this region stream salinities are also increasing rapidly [53].

3. Technical aspects

The relevant technical aspects of the salinity problem include understanding its cause, understanding the factors which affect its manifestation, and, among other approaches, determining what vegetation mixes can physically control the problem. The latter question is usually expressed as determining what species to plant, how to plant, where to plant, at what density and configurations to plant, and how much area to plant.

3.1. *Causes of salinisation*

It is a common observation throughout Australia that salinisation occurs as a result of rising water tables (unconfined aquifers) or rising potentiometric surfaces (semi-confined aquifers) [65, 35, 28, 43, 54, 48].

In dryland salinity, the rise in the water table follows a change in the water balance when the deep-rooted, perennial native vegetation is replaced by shallow-rooted, annual agricultural plants. The reduced water consumption by the agricultural plants leads to increased groundwater recharge and rising groundwater levels. It may take some years before groundwater levels reach the ground surface as is well demonstrated in the Lemon experimental catchment (Fig. 3). In the case of irrigation salinity, increased groundwater recharge occurs when irrigation water (or rainfall) passes below the root zone; i.e. exceeds the plants' requirements.

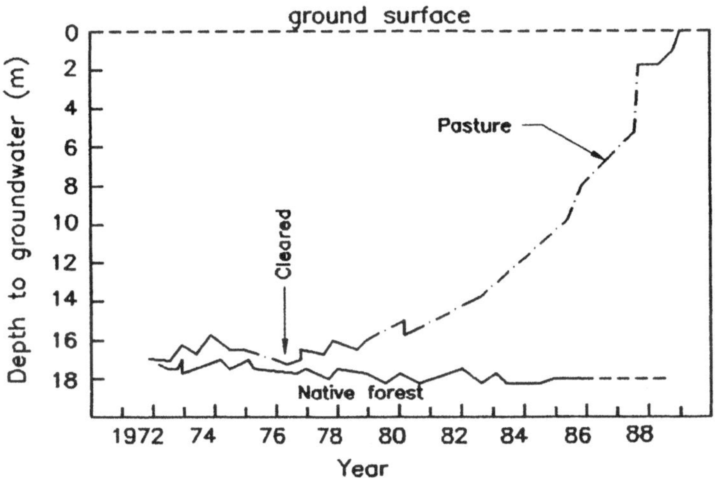

Fig. 3. Rise in groundwater level following clearing and pasture development in Lemon catchment, Western Australia [Source: 48].

In both dryland and irrigation salinity, a source of salts is necessary for a problem to emerge. The source in most situations is the salt accumulated in the unsaturated zone, commonly of oceanic origin.

3.2. *Factors affecting salinisation*

Many factors influence where, when and how much salinisation will occur. These include soil salt content, soil hydraulic properties, soil depth, geology, bedrock topography, dykes, nature and distribution of remnant vegetation, agricultural species and farming practices, annual rainfall, irrigation quantity and quality and irrigation scheduling. Some understanding of the role of these factors is important in determining best 'solutions' in each given situation. A recent example of the employment of strategic surveys in combination with modelling to describe and assess some of these important factors for tree planting is described by [40].

3.3. *Which species to plant*

Three important criteria are often quoted when addressing tree selection:

— adaptation to site conditions
— water use
— multiple use

Adaptation. This is the primary criterion to be satisfied. It includes adaptation to climate, soils, pests, diseases, fire, waterlogging and salinity.

Water use. Knowledge of the potential water use of species [25, 27] for a range of planting layouts and site conditions could allow trees to be selected to minimise the area required for tree planting. This is particularly important where there is competition with existing or other land uses.

Water use measurements have been carried out with parameters, ventilated chambers [15, 16] and the heat pulse technique [33, 39]. In some cases plantations have been estimated to transpire 3—4 times the rain falling on the plantations [24]. Tree water use, however, is sensitive to local conditions, including soil water and groundwater salinity [55]. To evaluate the water use potential of a wide range of species and provenances for various site conditions, climates and layouts would be an enormous task. The future direction of this work would be best directed at estimating the water use of a small sample of key species, climates, sites and layouts. A complementary approach is to determine the evapotranspiration characteristics of species as inputs to models which could quickly simulate water use in a wide range of situations.

Multiple use. This refers to the beneficial uses of trees other than for salinity control which may influence selection. These include commercial tree cropping for timber, pulp, firewood, fodder and other products; shelter and shade, wind and water erosion control, waterlogging control and aesthetics.

Although salinity control may be the prime objective, the attractiveness and feasibility of tree planting may depend strongly on some of these other uses.

The above are some of the criteria which guide the tree selection process. To help determine the stock from which to choose, [62] have listed about 40 native tree species with potential for use on recharge areas (areas where groundwater recharge is occurring) and discharge areas (areas where groundwater is discharging at the ground surface). Further information is now being compiled on the performance of these species in various climates and with varying degrees of waterlogging and salinity.

In Victoria extensive trials to identify tree species tolerant of saline conditions have been conducted by universities and government agencies under both glasshouse and field conditions [12, 38]. The importance of provenance variation became apparent during these trials, and there has been some investigation into this effect in trees of the *Eucalyptus camaldulensis* group [59].

In Western Australia more than 100 species of Australian woody plants have been screened for salt/waterlogging tolerance under glasshouse conditions [61]. At present 22 of the most tolerant species have been targeted to cover a range of climatic conditions throughout Australia (and overseas) and to serve a range of economic uses [36]. The performance of selected tree and shrub species in the field has been reported by [44].

More recently research has focused on micropropagation techniques for eucalypts, which offer the potential to produce clones of superior salt/waterlogging tolerant trees [60]. Substantial variation among individual seedlings had been observed in glasshouse screening trials, portending large improvements in the tree stock for saline seeps. This approach has progressed in Western Australia to the stage where field trials have demonstrated the superiority of clonal lines over unselected seedlings [56].

3.4. *How to plant*

On well drained recharge areas, normal tree planting methods are appropriate. These are explained in standard texts.

In or close to saline, ephemerally waterlogged discharge areas, special techniques are necessary for good establishment. First, well adapted species should be selected. Various site amelioration techniques have shown improved survival and growth performance. These include ridge (double ridge is preferred) mounding [45] combined with graded drainage [47], application of mulches, increasing seedling container size [46], ripping and herbicide spraying (the latter may not be necessary on harsh saline/waterlogged sites), and deep ripping when hardpans are present.

Direct seeding of trees has received close attention in Victoria [39] since it offers much reduced establishment costs where large numbers of trees are involved. The application of direct seeding for timberbelt establishment in dryland recharge areas is of particular interest. Research to improve the reliability of this practice is continuing in northern and western Victoria, although substantial success has already been achieved by effective weed control treatments at sowing. Direct seeding is in use operationally on farms in several States.

3.5. *Where to plant trees and how many*

In salinity control, the questions of where to plant trees and how many have been the most perplexing. This is because there is potentially a wide range of options when only part of the landscape is to be covered by trees. Also farming practicalities have a strong bearing on where and at what density trees can be planted. However, starting from the theoretical viewpoint, the reforestation strategy must meet one fundamental requirement:

> *(I) Annual evapotranspiration plus streamflow from the land must equal or exceed the rainfall.*

This assumes no groundwater leakage. This requirement is the hydrological balance which must be met if groundwater levels are not to rise. (I) usually necessitates, by implication, the presence of perennial vegetation in the lower landscape since trees on the upper landscape cannot control recharge on the lower lands. In some cases, however, recharge to a specific aquifer may occur primarily in upland areas.

To achieve (I), there are some ways in which trees can be used more efficiently. Firstly, as mentioned in 3.3, trees with high water use potential can be selected. Secondly, wider spaced trees develop higher leaf areas per unit ground cover [49] and their root systems can access larger soil volumes; these lead to higher water use per unit area planted. Thirdly, trees can be planted on areas of higher recharge and so more efficiently reduce recharge [63]. Fourthly, trees can be planted to intercept groundwater at the contact between recharge and discharge areas, thus avoiding severe discharge area conditions and areas high in the landscape where the depth to groundwater is substantial [21].

From the above it is apparent that the theoretical ingredients of a reforestation plan are fairly well known. The theory is constrained, however, by a number of practical considerations. The first constraint to be overcome is the availability of planting material which is adapted to the site (unless, for example, the species can be managed on a short rotation for commercial production, as with *Eucalyptus globulus*). If a number of species are well adapted, the choice can then be based on the economic optimum, that is the trade-off between any commercial value of the species, against its water use potential and consequent land area requirement. If, as is the case in Western Australia with *E. globulus*, the commercial potential is very high and the water use potential is at least moderately high, then a particular species can gain high preference. Optimising the commercial potential of tree planting may then govern such things as planting density and location. Choosing species on the basis of water use alone is, at the present time, ill-advised, because of the lack of comprehensive, good quality water use data.

A second constraint is agricultural manageability. Whereas wide-spaced trees may potentially be more efficient water users, they are also obstacles to farm machinery. As a consequence there has been a move towards tree belts which allow cropping to occur unhindered between the belts. Whilst this system has clear practical advantages, it poses questions in regard to meeting (I) such as optimum belt width, belt spacing and planting densities within belts. Recharge may also occur between belts. These questions require further research.

An alternative to systems such as wide-spaced and belt planting is planting areas of high recharge. Considerable study has been devoted to identifying potentially high recharge areas [62]. Among these techniques significant acclaim has been given to electromagnetic induction methods which identify high and low salt content profiles which, by inference, are low and high recharge areas respectively. Increasing and decreasing salinity profiles with depth have also been used to identify recharge and discharge or potential discharge areas [63]. However caution is required in pursuing this approach to the exclusion of other methods [22] since some high salt storage profiles may be permeated by fissures and channels which allow rapid recharge in saturated conditions [29]. Also lower landscape positions naturally accumulate water and often may go undetected as high recharge areas. Finally, in winter rainfall-dominated climates, evapotranspiration is relatively low when recharge occurs; in this case it may be more important to locate trees for optimum water extraction during summer rather than for recharge control.

The local climatic conditions also influence potential evaporation as noted in the modelling study of [41]. In hilly landscapes in particular, the solar radiation intensity and duration varies considerably with slope and aspect. This strongly affects evaporation, which in turn affects the ability of trees to control groundwater levels.

A final factor (but often one that needs to be addressed first) is the landowner's perceptions and desires. This can strongly influence species selected, where trees are planted and how many. If the government (on behalf of the general public) has a strong interest in the outcome, it may need to organise alternative approaches or purchase land to enable reforestation [13].

3.6. *Experimental observations of groundwater and salinity control by trees*

The most extensive set of observations derive from Western Australia [55, 10, 52, 3, 21]. From a number of experiments in a 700 mm annual rainfall region, a very close relationship was found to exist between groundwater levels beneath tree plantings and the amount of tree cover (Fig. 4). At the

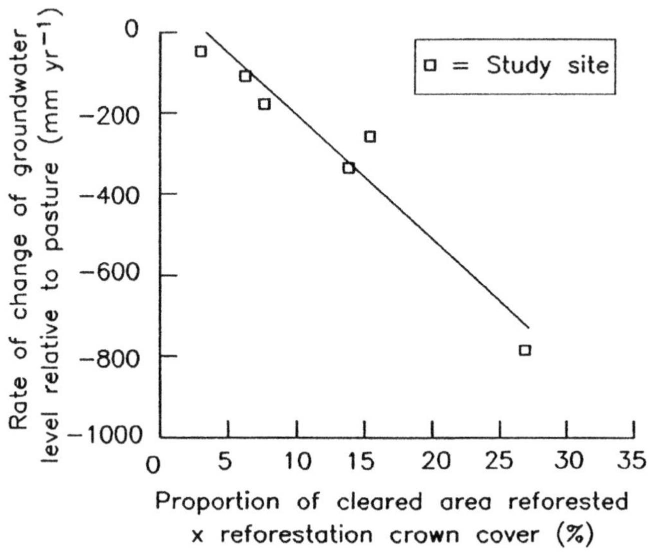

Fig. 4. Relationship between rate of change of groundwater level beneath tree plantings and amount of tree cover at six locations in Western Australia [Source: 50].

one extreme of extensive, high density, planting (covering 70% of cleared land, 625 stems ha^{-1}, 38% crown cover), groundwater levels were lowered on average 5.5 m relative to the ground surface across the site over a period of nine years [4]. However, for plantings of less than 15% of the cleared area, no such groundwater level reductions occurred. Such data, however, are site specific and should not be extrapolated to areas with different conditions.

There is also evidence of water table lowering by trees both in irrigation and dryland areas in Victoria. In a northern Victorian irrigation area an 8-year-old eucalypt plantation lowered the water table by 2 m or more, and reduced the piezometric head in the underlying aquifer by up to 1.5 m [26]. At several dryland sites, progressively greater and more rapid lowering of the water table was observed as planted trees (near saline seeps) grew to occupy the site more fully [39].

Although numerous examples of groundwater level lowering by trees have now been reported, there has been little data published on the consequent effects on groundwater and soil salinity. A study of groundwater salinity changes beneath four experimental tree planting sites in a 700 mm annual rainfall region in Western Australia indicated a salinity decline of 11% over the first seven years since planting [55]. At a lower rainfall site (500 mm) an agroforestry plantation lowered the groundwater level by up to 1.9 m but the groundwater salinity beneath trees generally increased [55]. At the extensive reforestation site [4] there was evidence of salt redistribution in the soil profile but a consistent picture did not emerge. In the lower rainfall site, the soil salinity decreased from 0—0.75 m, increased from 0.75—2.25 m and remained approximately the same below this depth [55]. In another study with 62% pine and 7% eucalypt reforestation (the valley floor was left under pasture and 20% of the catchment retained native forest), the salt export from the catchment decreased by 90% whilst the stream salinity increased by 25% over the period 1978—86 [14].

Salt dynamics in relation to tree planting requires further careful research, particularly in lower rainfall and irrigation areas.

3.7. *Use of models in tree planting*

Since experimental data is constrained to local conditions and a restricted climatic period, there is a high potential to make use of modelling in a wide variety of situations. Three modelling attempts have been reported.

The first was a simple water balance model which can estimate tree area planting requirements, rates of water table lowering etc. based on annual water use data and areas occupied by the different vegetation types [51]. The second was a groundwater model in which the vegetation modifies the recharge/discharge functions [51]. The third was a detailed biophysical model (TOPOG) which attempts to simulate all the key processes. An early version of this model has been used to determine the optimum location of tree planting to reduce the area of a saline seep [41]. A more sophisticated version which includes a permanent groundwater aquifer as well as the shallow perched aquifer is currently being developed. The models can be partially validated on the experimental data and then used as a tool in a much wider range of conditions.

4. Economic and financial aspects

4.1. *Benefit-cost analyses*

A full financial and economic evaluation of reforestation would require analysis of all the benefits and costs. Benefits accrue to the landowner and community as returns from wood production, shade and shelter for domestic animals, habitat for wild animals and birds, wind and water erosion control, farm business diversification, improved farm landscape, and control of groundwater, waterlogging, nutrient discharge and salinity. It has also been claimed that significant increases in crop/pasture production are possible due to improved microclimatic and soil conditions due to sheltering. It has been demonstrated [11] that up to 20% of an undulating treeless farm could be planted to tree belts without loss of agricultural production. Improved landscape aesthetics further contributed to an increase in land values of Victorian farmers participating in the Potter Farmland Plan [57]. Costs include loss of some agriculturally productive land; land preparation, planting material, planting, fencing, management of stands (pruning, thinning as necessary) and logging and transport (if harvested).

In the benefit-cost analyses conducted to date, most of the costs indicated above have been included, whilst the benefits have generally been limited to commercial produce (forest and agricultural). In Victoria economic analyses of salinity control techniques have generally shown tree growing to be expensive, largely due to an expected reduction in agricultural income and the presumed inability of trees on many sites to yield commercial products [18, 32, 42]. The validity of these analyses has been somewhat flawed by a lack of reliable information on the interaction between trees and agriculture, and the lack of a methodology for evaluating the salinity control benefits of tree establishment projects [39].

The advent of agroforestry raised the possibility of increased productivity as well as better land protection by trees and agriculture in a co-operative rather than competitive relationship. Early results from agroforestry research in Victoria were described by [2, 30]. A detailed analysis for a pine-pasture agroforest in Western Australia [1, 37] demonstrated a net overall financial gain of 40% compared to agriculture alone. A simulation model for the economic evaluation of agroforestry systems has recently been developed [31] and will provide a valuable tool for economic analysis of tree growing options into the future [39]. The structure of the model allows for progressive updating of cost, growth and yield data as more information becomes available from research and experience.

4.2. *Commercial wood production*

Probably the most significant development in catchment restoration in the last 5 years has been the recognition of the commercial potential of tree plantations on farmland [5, 34]. This situation has arisen due to several factors:

- environmental pressure to reduce logging in native forests;
- economic pressure to reduce the level of timber imports;
- environmental pressure to cease importing timber where its felling is degrading important global natural resources or contributing to global climate change;
- a viable, stable global pulpwood market;
- recognition that tree planting can help solve many land and water degradation problems (including salinity);
- recognition that commercially valuable species can form a major component of most reforestation schemes.

These factors, together with strong community support for landcare initiatives, are creating a more positive environment for commercial tree planting. However, there is a number of constraints and obstacles:

- tree planting is generally only commercially viable above 600 mm annual rainfall;
- tree planting is only commercial where there is an adequate timber/pulpwood production infrastructure and transport costs are not prohibitive;
- optimum or commercial yields may only be available under specific soil conditions;
- long crop rotation periods can mean an uneven distribution of cost and returns and increased risks [7];
- negative sentiments towards tree planting and monocultures on agricultural land;
- need for additional, often non-commercial species to be planted to achieve reclamation objectives.

To overcome some of these constraints, forestry sharefarming was developed in Western Australia in 1987 [55] and timberbelt sharefarming in 1990 [6]. In these schemes the State or private investor(s) and the farmer agree to share the establishment and management costs of the tree planting on the farmer's land and to share the harvest revenue. The timberbelt sharefarming is a variation in which timberbelts rather than block plantations are used. A timberbelt is a stand of trees at least several rows wide and of any length or shape. This flexibility has proven attractive to farmers. Belts are integrated within the farm plan and designed to give optimum wood production and other benefits. A Timberbelt Sharefarming Scheme for south-west Western Australia has been found to remain commercially viable with 20% of plantings devoted to non-commercial species on salt-affected land.

4.3. *Sub-commercial production*

Sub-commercial wood production occurs where trees are harvested in some way to provide benefits which would only partly offset the costs. This is likely to be particularly important in lower rainfall areas. It may be where a less productive, more salt tolerant tree species is used, or where trees are logged for on-farm uses such as fence posts and firewood. Other products with income generating potential include flowers, essential oils, seed and fodder.

5. Socio-political factors

5.1. *Political factors*

Political factors can have a strong bearing on any major tree planting programme. Government may either be pressured by the community or interest groups to take action or may recognise the need to intervene for the public good. An example of intervention occurred in Western Australia when in 1976 and 1978 the government legislated to control tree clearing in important south-western catchments with marginal stream salinity. This was followed in 1979 by the commencement of a major government-funded reforestation programme on the Wellington Dam catchment, the largest impounded water resource in the south-west. The reforestation programme involved the purchase of farms by the Water Authority as they became available and the exchange of productive land for saltland. Thirty per cent reforestation was targeted, covering the valley floor and lower slopes. The total cost of the programme to 1990, which has achieved 6,200 ha of dense tree planting, is $15M [13].

Strong political support for land restoration through tree planting has been expressed and various national tree planting schemes initiated (e.g. the Commonwealth's One Billion Trees Program, National Afforestation Program and National Soil Conservation Program). However government funding has been generally little more than 'seeding' money. The early 'interventionist' approach has lost favour, partly through lack of finance and partly by the wish to place implementation of the solution back with the landowners and local communities. Government has also become more tentative in the commercial plantations where private investors could equally participate.

5.2. *Social factors*

Social factors relate to the social structure that has developed in farming communities around existing farming practices, their culture, attitudes and lifestyle; and the impacts of change required for catchment rehabilitation. Pioneering agricultural development of the land involved clearing vast tracts of native vegetation where trees were seen as obstacles to agriculture. Following clearing, regrowth control was often necessary for several years. The cleared land varied in productivity and was often marginal until the advent of high solubility and trace element fertilizers. Clearing boomed until the late 1970s, strongly supported by governments, and is still active in many parts of Australia. In more recent times land and water degradation has emerged on a scale to warrant national concern and the task of changing entrenched farming practices and attitudes that lead to degradation of land and water resources has commenced. Throughout Australia the National Soil Conservation Programme has emphasised the mobilising of local community forces. A broad awareness of the potential role of trees has arisen and research has gone some way towards developing agroforestry solutions. Nevertheless there is a long way to go in developing sustainable land uses, in gaining confidence in agroforestry solutions and in convincing a cautious and conservative rural public to adopt these solutions.

6. Administrative and management approaches

6.1. *Administrative approaches*

The Australian States are in various stages of addressing the salinity problem, both physically and administratively. Whilst the earliest administrative actions occurred in Western Australia, the most systematic and sophisticated administrative approach is in Victoria [39]. Here in 1982 a Parliamentary Committee was established to conduct an inquiry into land and water salinity. The Committee's final report in 1984 [42] recommended the creation of a central salinity control agency (the Salinity Bureau) to co-ordinate and oversee planning and implementation by government bodies and community groups. Increased State and Federal funding was made available to support a range of initiatives.

Following extensive community consultation, a State Salinity Management Strategy was released by the Bureau in 1988 [23]. The strategy set out the government's policy, 10-year targets and plans for future action in eight programme areas:

— Regional planning and co-ordination
— Farm improvement and assistance
— Revegetation
— Improved water use
— Drainage and salt disposal
— Environmental protection and improvement
— Education and participation
— Research and investigation.

For implementation of the programmes, the State was divided into nine salinity control regions based on major catchments. The Strategy required preparation of a salinity plan for each region, and detailed salinity management plans for sub-regions with significant problems. The plans are prepared by working groups of representatives from the regional community, with assistance from consultants and government agencies as needed. Regional salinity forums have also been established in a number of regions, bringing together community representatives to co-ordinate the salinity control efforts of government agencies, farmers and private organisations, and to advise the government on regional salinity control priorities. A description of the progress to date is given in [39].

Administration of salinity programmes in most other States has become part of the Commonwealth Government's Decade of Land Care. In this approach, Land Conservation District Committees (or their equivalent) are formed by local communities to address local land management problems including salinity. The State Government provides technical support but leaves the essential decision making in the hands of the committees. Each

State is also producing a Decade of Land Care Plan which aims to provide a broad framework to guide governments and communities.

6.2. *Management approaches*

Management approaches are not necessarily distinguishable from administrative approaches but are used here to describe the philosophy of managing the problem rather than the administrative process.

Management of land in Australia is now almost universally considered within the framework of Total Catchment Management or Integrated Catchment Management (ICM). These concepts are based on the realisation that landscape attributes and values can be better managed within the natural boundaries of hydrological (surface water and groundwater) catchments rather than alternative administrative boundaries. This recognises the role of water as both a major transporter of materials (e.g. salt) within and from the landscape and as an indicator of the health of a catchment. An additional advantage of the catchment approach is that a catchment can be broken down into subcatchments of ever decreasing size whose individual outputs can be combined to determine downstream impacts. As such this approach aids understanding and is highly amenable to computer modelling.

In regard to tree planting for salinity control, the catchment approach is essential since it is the basis for determining the vegetation requirements (type, density, distribution, areas) to control groundwater levels. ICM, though, is more than this, because it also questions the impacts that 'solutions' such as tree planting have on the other values of the catchment. ICM thus requires a broader view of land management and necessitates a careful and early evaluation of objectives (e.g. production, development, protection, rehabilitation etc.) for the catchment.

7. The future

The contributing papers and this overview identified some specific requirements for future study (Table 2). The following provides some comments on where the author sees future needs and progress.

Table 2. Specific requirements for future study.

Salinisation

1. Further study of salt accumulation beneath trees planted above saline groundwater.
2. Continue research into the processes underlying the salinity problem to enhance understanding and permit the development of more effective control measures.
3. Monitoring of baseline salinity levels and water table depths and areas so that it can be determined if salinity is increasing, at what rate, and if work already established is reducing critical salinity levels.
4. Provide more information on water and salt movement in shallow aquifers.

Table 2 (Continued)

Technical aspects of the tree planting approach

5. Continued identification and development of salt/waterlogging tolerant genetic material.
6. Additional evaluation of tree species for salt tolerance at the provenance level.
7. Testing of greater range of species and provenances of species in different locations.
8. Water use and physiology of individual tree species.
9. Investigation of high water use vegetation systems and their productivity and profitability.
10. Continued development of site preparation and propagation technology to enhance the survival and development of the plants.
11. More study of direct seeding techniques.
12. Study of the effect of excluding stock on the regeneration of trees in different parts of the landscape.
13. The relationship between tree density and groundwater accession rates in dryland areas.
14. Determination of the relationship of discharge area plantings to water table and soil and groundwater salinity levels.
15. Development of models and site surveys to test tree planting strategies and develop vegetation plans.
16. Develop commercial aspects of tree planting to improve the cost effectiveness of incorporating trees into farm management.
17. Determination of suitable tree species and optimum stocking rates to maximise crop/pasture production through tree density/pasture interaction studies.
18. Identify and develop agroforestry activities with commercial potential for low rainfall areas.

Economics and financial aspects

19. The economic effectiveness of trees as a salinity control measure.
20. Quantify the other benefits of planting trees for salinity control.
21. Compare other salinity control strategies with tree planting.
22. Financial planning practices for tree plantations.

Socio-political factors

23. Identify the needs, expectations and inhibitions of farmers with respect to their salinity problems and the incorporation of trees into farm management.
24. Determine how to increase awareness and change the attitudes to salinity of landowners, communities and government agencies.
25. Minimise obstacles to adoption and implementation of ICM principles.

Administration and management

26. Encouragement of collaborative studies combining tree, agricultural (crop and animal) and water disciplines. Ultimately, all facets need to be integrated into new farm management systems, resulting in strategic use of trees.
27. Set long term goals for salinity control and tree planting.
28. Encourage the adoption of integrated catchment management principles at regional levels.
29. Provision of information on tree planting, salinity and interactions, especially to farmers.
30. Establishment of community structures with ability and desire to control and manage change and take ownership of problems.
31. Provide clear strategies for government and community programmes.

Salinisation

Most States predict a worsening and expansion of dryland and irrigation salinity over the next several decades. As long as this continues, salinisation and its control will remain a major community issue. Further research will be required to gain a better understanding of local salinisation mechanisms and to determine the best solutions for particular localities.

Technical aspects of the tree planting approach

Most aspects discussed in this paper require continued research. Sufficient data are now available to demonstrate the potential of tree planting as a salinity control measure. More work, however, is required to answer the basic questions of which planting material, how to plant, where to plant, at what density and configurations, and how much area — across a range of conditions. Further advances will be made in tree technology such as selection, cloning, hybridisation, productivity. Site amelioration for improved establishment will provide further gains. Modelling has yet to make a major contribution, but substantial steps forward in computer power, cost, software and biophysical input data herald a productive future. Strategic surveys such as electromagnetic, soil, geology, topography, combined with modelling, will increasingly become a necessary preliminary to developing the vegetation strategy for salinity control. Field trials will continue to be necessary for validation of models and as demonstrations. In some hydrogeologic systems, there is a need for more study of possible salt accumulation beneath trees, and its effects.

Economic and financial aspects

Commercial tree planting on farmland has great potential to assist salt reclamation but this option is generally viewed as only tenable in higher rainfall areas. Economic analyses, though, in any situation will need to include a fuller set of benefits (which in turn require better input data) from trees if they are to be credible. This may require innovative economic techniques.

Tree planting will require not only sound farm planning but also wise financial planning. Tree planting may be a long term investment in the viability of agriculture and the landscape but should not be seen as a quick fix or panacea for all problems. Economic and financial comparisons between alternatives (e.g. saltland agronomy versus tree planting) will need to be considered.

Socio-political aspects

The environment and trees are currently high on the political and community agenda, but this may not always be the case. Whilst advantage should be taken of the current political climate, a longer-term view of land management will be necessary. Setting goals for the future is always useful, and exploring alternative futures, at least hypothetically, should help identify where the landowners and community want to head. Testing new concepts, better communication, and closer interaction between the community and government and its agencies will all pay dividends.

The development of positive sentiments towards trees in the rural community is already occurring and will need to be fostered further. With more 'power' being placed in local land management groups, better and more readily accessible information is necessary, along with the harder questioning of the providers of information. Change is often acceptable to people, but the ability of the individual and community to control and manage change for everyone's benefit will be crucial.

Administrative and management approaches
Administrative arrangements vary from State to State and with the magnitude of the problem. Nevertheless a clear government strategy and well co-ordinated and participatory programmes will be increasingly essential.

Integrated or Total Catchment Management has gained prominence throughout Australia. A clear relationship between detailed farm planning and ICM exists. However, there are clearly obstacles to its successful implementation, such as the potential for conflict between vested interests, different views and emphases on catchment objectives, and administrative difficulties with natural and man-made boundaries. Despite these problems the potential benefits are high.

8. Concluding remarks

The knowledge and understanding of the potential of tree planting to control salinity has advanced greatly in the last decade. Small test sites, demonstrations and farmer plantings have become abundant in some areas. Major tree planting programmes, however, are struggling in their infancy. The main barriers are cost, uncertainty and attitudes. Tree planting, though, should not be adopted with over-zealous fervour, but should be progressively tested, researched and analysed, always maximising feedback and refining. A place for trees in the landscape is essential, and our knowledge of the role of trees in controlling salinity is improving, but there are many and varied factors which govern realisation.

Acknowledgements

I am grateful to the authors of the contributing papers for use of their material and to Mr. P. Scott, Mr. R. George, Ms. W. Boddington, Mr. J. Bartle, Mr. N. Pettit and Mr. M. Bari for their comments on this paper.

References

1. Anderson GW, Moore RM and Jenkins PJ (1988) The integration of pasture, livestock and widely-spaced pine in south west Western Australia. Agroforestry Systems 6: 195—211
2. Baldwin P, Stewart H, Bird P, Hamer W, Flinn D, Cumming K and Connor D (1988) Establishment of an agroforestry research network in Victoria, Australia (1983—85). Agroforestry Systems 6: 213—233
3. Bari MA and Schofield NJ (1991) Effects of agroforestry-pasture associations on groundwater level and salinity. Agroforestry Systems 16: 13—31
4. Bari MA and Schofield NJ (1991) Lowering of a shallow, saline water table by extensive eucalypt reforestation. J Hydrol 133: 273—291
5. Bartle JR (1989) Eucalypt pulpwood — a potential new crop in the lower south-west. In: Arkell PT, ed, Trees — Everybody Profits. Proc Seminar, Como, W.A., Aust Inst Agric Sci, Occ Publ No 45: 41—50
6. Bartle JR (1991) Tree crops for profit and land improvement. WA Journal of Agric 32: 11—17
7. Bartle JR and Shea S (1989) Development of a Pulpwood Cropping Industry in Western Australia. Land and Water Research News 1: 5—9
8. Bastick CH (1991) The role of trees in controlling salinity in Tasmania. In: The Role of Trees in Sustainable Agriculture, pp 77—80. Proc Nat Conf, Sept 30—Oct 3, 1991, Albury, Australia
9. Beetson TB and Gordon IJ (1991) Role of trees in alleviating secondary salinity: the current position in Queensland. In: The Role of Trees in Sustainable Agriculture, pp 51—57. Proc Nat Conf Sept 30—Oct 3, 1991, Albury, Australia
10. Bell RW, Schofield NJ, Loh IC and Bari M (1990) Groundwater response to reforestation in the Darling Range of Western Australia. J Hydrol 115: 297—317
11. Bird R (1988) Financial gains of trees on farms through shelter. Proceedings of Bicentennial Conference, Albury-Wodonga
12. Blake T (1981) Salt tolerance of eucalypt species grown in saline solution culture. Aust For Res 11: 179—183
13. Boddington W, Kikiros G and Schofield NJ (1991) Stream salinity management in south west Western Australia: past achievements and future directions. Aust Water & Wastewater Assn Conf, Perth, W Australia, March 18—22, 1991
14. Borg H, Bell RW and Loh IC (1988) Streamflow and stream salinity in a small water supply catchment in southwest Western Australia after reforestation. J Hydrol 103: 323—333
15. Borg H and Giles R (1988) Transpiration by Trees — A Review of Recent Work Conducted in Western Australia. Surface Water Branch, Water Authority of Western Australia, Rep No WS26, 53 pp
16. Clifton C, Morris J, Trebilcock B and Miles P (1991) Developing Agronomic Options for the Control of Dryland Salinity: II. Water Use by Eucalypt Trees Growing on a Rocky Ridge at Burke's Flat. Dept of Conservation and Environment Tech Rep, Melbourne
17. Cooper MJ and Richley LR (1989) Land Degradation in Tasmania. Dept Primary Industry, Fisheries and Energy, Tasmania, 15 pp
18. Cope F (1958) Catchment Salting in Victoria. Soil Conservation Authority of Victoria, Tech Comm 1, 88 pp
19. Dooley T (1991) The role of trees in controlling dryland salinity — a South Australian overview. In: The Role of Trees in Sustainable Agriculture, pp 45—49. Proc Nat Conf, Sept 30—Oct 3, 1991, Albury, Australia
20. George RJ (1990a) The 1989 Saltland Survey, Journal of Agriculture of Western Australia 31: 159—66
21. George RJ (1990b) Reclaiming sandplain seeps by intercepting perched groundwater with eucalypts. Land Degrad & Rehab 2: 13—25

22. George RJ, McFarlane DJ and Lewis MF (1991) A review of recharge and implications for management in saline agriculture catchments, Western Australia, pp 193—197. International Hydrology & Water Resources Symposium, Aust Inst Engs, Perth, Oct 2—4, 1991
23. Government of Victoria (1988) Salt Action: Joint Action — Victoria's Strategy for Managing Land and Water Salinity. Government Printer, Melbourne, 54 pp
24. Greenwood EAN, Klein L, Beresford JD and Watson GD (1985) Differences in annual evaporation between grazed pasture and *Eucalyptus* species in plantations on a saline farm catchment. J Hydrol 78: 261—278
25. Greenwood EAN (1986) Water use by trees and shrubs for lowering saline groundwater. Reclam Reveg Res 5: 423—434
26. Heuperman A, Stewart H and Wildes R (1984) The effect of eucalypts on water tables in an irrigation area of northern Victoria. Rural Water Commission of Victoria, Water Talk 52: 4—8
27. Hookey GR, Loh IC and Bartle JR (1987) Water Use of Eucalypts Above Saline Ground Water. Water Authority of Western Australia, Water Resources Directorate, Rep No WH32, 76 pp
28. Jenkin J (1981) Terrain, groundwater and secondary salinity in Victoria, Australia. Agric Water Manage 4: 143—171
29. Johnston CD (1987) Preferred water flow and localised recharge in a variable regolith. J Hydrol 94: 129—142
30. Kellas J, Bird P, Bienvenu F, Cumming K, Hall R, Lyons A and Morgan M (1989) Agroforestry in Victoria: Progress in Research and Extension. Dept of Conservation, Forests and Lands/Dept of Agriculture, Melbourne, 35 pp
31. Loane W (1990) Farm Tree: Simulation Model for Economic Evaluation of Agroforestry and Trees on Farms. Introductory manual. Joint Agroforestry Management Committee, Victoria, 17 pp
32. Lyle C (1984) The role of on-farm methods in controlling salinity in the Shepparton and Kerang irrigation regions. In: Dwyer Leslie Pty Ltd, eds, Salinity Control in Northern Victoria — a Strategic Study for the Salinity Committee of the Victorian Parliament. Government printer, Melbourne, Working Paper D
33. Marshall JK and Bell DT (1991) Water uptake of clones: merging technologies bring funding success. Land & Water Research News 10: 13—16
34. Mattinson B, Morrison D and Eckersley P (1989) Economics of tree farming in the south-west of Western Australia. In: Arkell PT, ed, Trees — Everybody Profits, pp 70—83. Occasional Publication 45, Aust Inst Agric Sci
35. Macumber P (1969) Interrelationship between physiography, hydrology, sedimentation and salinization of the Loddon River plains, Australia. J Hydrol 7: 39—57
36. McComb JA, Bennett IJ, Bell DT and van der Moezel PG (1989) Biotechnology enhances utilisation of Australian woody species for pulp fuel and land rehabilitation. Aust J Biotechnol 3: 297—301.
37. Moore R (1989) Agroforestry with widely spaced trees. In: Arkell PT, ed, Trees — Everybody Profits, pp 55—65. Occasional Publication 45, Aust Inst Agric Sci
38. Morris J (1984) Establishment of trees and shrubs on a saline site using drip irrigation. Aust For 47: 210—217
39. Morris JD (1991) The use of trees in salinity control in Victoria, Australia: progress and potential. In: The Role of Trees in Sustainable Agriculture, pp 63—75. Proc Nat Conf, Sept 30—Oct 3, 1991, Albury, Australia
40. Nicholson AT (1991) Catchment planning and use of trees in salinity control: Yahoo Peaks dryland salinity control demonstration. In: The Role of Trees in Sustainable Agriculture, pp 59—62. Proc Nat Conf, Sept 30—Oct 3, 1991, Albury, Australia
41. O'Loughlin EM, Short DL and Dawes WR (1989) Modelling the hydrological response of catchments to land use change. Hydrology and Water Resources Symposium 1989, Univ of Canterbury, Christchurch, NZ

42. Parliament of Victoria (1984) Salt of the Earth: Final Report on the Causes, Effects and Control of Land and River Salinity in Victoria. Government Printer, Melbourne, 322 pp
43. Peck AJ and Williamson DR (1987) Effects of forest clearing on groundwater. J Hydrol 94: 47—66
44. Pettit NE and Ritson P (1991) Performance of Selected Tree and Shrub Species Grown for Stream Salinity Control in the Wellington Reservoir Catchment. Water Authority of Western Australia, Rep No WS 78, 60 pp
45. Ritson P and Pettit NE (1992) Double-ridge mounds improve tree establishment in saline seeps. For Ecol & Manag 48: 89—98
46. Ritson P and Pettit NE (1989) Seedling containers and mulching for tree establishment in saline seeps. Land and Water Research News 3: 16—17
47. Ritson P and Pettit NE (1991) Drainage of saline seeps to improve tree establishment. Land and Water Research News 8: 28—30
48. Ruprecht JK and Schofield NJ (1991) Effects of partial deforestation on hydrology and salinity in high salt storage landscapes. I. Extensive block clearing. J Hydrol 129: 19—38
49. Ruprecht JK and Schofield NJ (1991) Effects of partial deforestation on hydrology and salinity in high salt storage landscapes. II. Strip, soils and parkland clearing. J Hydrol 129: 39—55
50. Schofield NJ (1990a) Effects of trees on saline groundwater tables. In: Scott PR, ed, Agroforestry — Integration of Trees into the Agricultural Landscape, pp 10—30. Div Resour Manag, WA Dept Agric, Tech Rep No 102
51. Schofield NJ (1990b) Determining reforestation area and distribution for salinity control. Hydrol Sci J 35: 1—19
52. Schofield NJ and Bari MA (1990) Valley reforestation to lower saline groundwater tables: results from Stene's Farm, Western Australia. Aust J Soil Res 29: 635—50
53. Schofield NJ and Ruprecht JK (1989) Regional analysis of stream salinisation in south-west Western Australia. J Hydrol 112: 19—39
54. Schofield NJ, Ruprecht JK and Loh IC (1988) The Impact of Agricultural Development on the Salinity of Surface Water Resources of South-West Western Australia. Report to the Steering Committee for Research on Land Use and Water Supply. Water Authority of Western Australia, Rep. No. WS27, 83 pp
55. Schofield NJ, Loh IC, Scott PR, Bartle JR, Ritson P, Bell RW, Borg H, Anson B and Moore R (1989) Vegetation Strategies to Reduce Stream Salinities of Water Resource Catchments in South-West Western Australia. Report to the Steering Committee for Research on Land Use and Water Supply. Water Authority of Western Australia, Rep. No. WS33, 98 pp
56. Schofield NJ, Bari M, Bell DT, Boddington WJ, George RJ and Pettit NE (1991) The role of trees in land and stream salinity control in Western Australia. In: The Role of Trees in Sustainable Agriculture, pp 21—43. Proc Nat Conf, Sept 30—Oct 3, 1991, Albury, Australia
57. Seddon G (1989) Revegetation — bringing back our Australian landscape. In: Arkell PT, ed, Trees — Everybody Profits, pp 8—11. Occasional Publication 45, Aust Inst of Agric Sci
58. Soil Conservation Service of New South Wales: Dryland Salinity, 6 pp
59. Thomson L (1988) Salt tolerance in *Eucalyptus camaldulensis* and Related Species. PhD Thesis, University of Melbourne, 307 pp
60. van der Moezel PG, Bell DT, Pearce-Pinto GVN and Kabay ED (1989) Clonal trees for saline areas throughout Australia. Research into Rural Tree Decline, Annual Newsletter: 13—16
61. van der Moezel PG and Bell DT (1990) Saltland reclamation: selection of superior Australian genotypes for discharge sites. Proc Ecol Soc Aust 16: 545—549
62. Walker J, Williams B, Hatton T, O'Loughlin E and Jupp D (1991) Use of trees for dryland salinity control. In: The Role of Trees in Sustainable Agriculture, pp 81—94. Proc Nat Conf, Sept 30—Oct 3, 1991, Albury, Australia

63. Williams BG and Somsri Arunin (1990) Inferring recharge/discharge areas from multi-frequency electromagnetic induction measurements. Technical Memoranda 90/11, CSIRO, Canberra
64. Williamson DR (1990) Salinity — an old environmental problem. In: Year Book Australia 1990, pp 202—211. Canberra, Aust Gov Publ Serv
65. Wood WE (1924) Increase of salt in soil and streams following the destruction of the native vegetation. J Royal Soc WA 10: 35—47
66. Working Party on Dryland Salting in Australia (1982) Salting of Non-Irrigated Land in Australia. Soil Conservation Authority, Victoria

Agroforestry Systems **20**: 25–58, 1992.

Alternative products from trees and shrubs to 'The Role of Trees in Sustainable Agriculture' conference, Albury, September 30th–October 3rd 1991

A. BOUTLAND[1], M. ROBINSON[1,*], J. FIELD[1,*], M. SCHELTEMA[2], B. HAWKINS[3], P. BULMAN[4], J. SCARVELIS[4], B. MASON[4], M. WALLACE[4], M. CASTLEY[5], P. A. RYAN[6], R. T. HENDERSON[6], G. B. APPLEGATE[6], C. LANSDOWN[7] and R. REID[8]

[1] *Forestry Dept., A.N.U., GPO Box 4, Canberra 2601, Australia;* [2] *Greening Australia, W.A.;* [3] *Forestry Commission, N.S.W.;* [4] *Woods & Forests Department, S.A.;* [5] *Private Forestry Division, Tasmania;* [6] *Queensland Forest Service;* [7] *Department Conservation & Environment, Victoria;* [8] *Forestry School, Creswick, Melbourne University*

Introduction

The potential role for incorporating the production of alternative tree and shrub products into normal farm management is an area which has received little attention to date. Despite a 40,000 year history of exploitation by Aborigines very few indigenous plant species are commercially exploited in Australia [Cribb and Cribb, 1981], apart from those for timber and fodder uses. Equally neglected has been the production of alternative products from non-indigenous species. In this paper, we examine the position by focussing on vegetation products, apart from timber and fodder which are potential sources of revenue. This paper reviews the current economic significance of many of the present alternative tree and shrub industries in Australia, the research and development (or lack of) associated with these products and finally, the potential role they could play in enhancing on-farm income.

Alternative tree and shrub products which are significant in Australia include: essential oils, honey and pollen, wildflowers and tree foliage, broom-bush, nut and other tree crops, sandalwood, craftwood, seed, christmas trees, drugs, tannins, gums and resins, cane, cork and charcoal. Many of these existing industries rely on exploiting natural resources, but increasingly plantations of desirable species are being established. Alternative tree and shrub products may be produced Australia wide (e.g. honey production), but in many cases products are restricted to the vegetation communities of particular climatic zones (Queensland rainforests have been a rich source of medicinal plants in Australia, [e.g. Maiden, 1889]).

This review examines the most important and potentially important alternative tree and shrub product industries in Australia, the plant species they depend upon and the agro-climatic region where they are based. Where possible, estimates of the economic viability of on-farm production are included. Based on these estimates, industries have been categorized as currently viable, potentially viable or possibly viable. At present it appears that many alternative tree and shrub product industries have little potential for enhancing on-farm income, unless there is a pre-existing resource. The cost of establishing a plantation resource specifically for minor forest production in many cases exceeds the expected return especially if costly inputs like irrigation are required. However, in many areas of Australia where there is an urgent need to revegetate, it should be possible to integrate trees or shrubs for alternative products into plantings for shade, shelter or erosion control.

Currently viable industries

Honey and pollen production

Production, value and markets

The honey and pollen industries of Australia are important in most regions and climatic zones. Much of the industry is based on native species (predominantly eucalypts), but exotic pasture and weed species also provide a large nectar supply for the industry.

In 1989—90 Australia's 836 apiarists produced some 27,000 tonnes of honey [Australian Honey Board, 1990]. The general trend is towards increasing production, although variations due to seasonal differences can be significant. New South Wales traditionally produces the most honey, with the Northern Territory the only non-producer. See table below.

Honey Production by States 1988—89.

State	Tonnes
New South Wales	9,619
Queensland	4,948
Victoria	2,886
South Australia	2,284
Western Australia	2,182
Tasmania	700

Source: Australian Honey Board, 1990.

Australia's beekeeping industry also produced 530 tonnes of beeswax in 1988—89 with a gross value of $1.9 million. The gross value of honey produced in 1989—90 was $29.6 million. More than half of Australia's honey crop is exported in bulk, earning $16.4 million in 1989—90. Most is exported to the United Kingdom, Germany and the USA [Australian Honey Board, 1990].

The beekeeping industry is important not only for the direct production of high quality honey, but also indirectly through the incidental pollination of many agricultural crops. For example, the value of the honey and beeswax industry to Tasmania is conservatively put at $1.3 million [Bureau of Statistics figures, 1986/87], not including the value of pollination to the State's $40 million crops industry [Tasmania Beekeepers Association, 1990].

Plant species and geographic location

Virtually all commercial honey production in Australia is from hives that are moved from one source of pollens and nectar to another. This is economically necessary in Australia because extremely variable rainfall and climatic conditions affect not only the flowering and budding patterns of the flora, but also the pollen and nectar yields [Clemson, 1985].

There are over 100 pollen and nectar bearing plants that are favoured by domestic bees, including Eucalypt species, acacias, *Callistemon, Banksia, Grevillea, Leptospermum*, blackberries (*Rubus fruticosus*), willows (*Salix* spp.), false tree lucerne (*Chamaecytisus palmensis*), various fruit and nut trees, some field crops which are amenable for on-farm plantings and some weed species. Most Australian honey is produced from eucalypts, although 70% of Tasmania's honey crop is produced from the nectar of leatherwood (*Eucryphia lucida*). The natural habitat of leatherwood is within temperate rainforests contained largely in State Forests and other Crown Lands of Tasmania's west coast.

In Western Australia most hives are located in the south-west region, in higher rainfall areas. The industry is heavily dependant on native vegetation. Two thirds of production originates from *Eucalyptus* species, the dominant being marri (*E. calophylla*), and the rest is produced from coastal heathlands and banksia woodlands [see Scheltema, 1991].

The major sources of nectar and pollen in Victoria are *Eucalyptus spp.* (mainly boxes) *Banksia* spp., tea-tree, ground cover crops such as clover and lucerne; noxious weeds such as blackberry, thistles and Patterson's Curse (*Echium plantaginaem*) are also important.

Problems and research

The major problem facing Australia's beekeeping industry is the viability of production to individual apiarists. For example in Western Australia the cost of production is $1.49/kg while the return to producers is only 89c/kg [Western Australian Department of Agriculture, 1989/90]. The high costs of production to the commercial producers are due primarily to the high transportation costs associated with the movement of hives to follow 'nectar flows'. Secondly, in many regions the decreasing native vegetation resource, due in part to clearing for agriculture, has significantly reduced the number of available nectar sources. In Western Australia these problems are exacerbated by short production rotations of Karri and competition for Banksia flowers with wildflower pickers. Thus, one aspect of research currently underway in W.A. is the identification of natural honey production areas and an evaluation of the relative importance of crown and public lands. A further important area of research is Queen Bee breeding where W.A. is an acknowledged world leader. Areas of likely future research include further work on bee breeding and management of hives to increase production [see Scheltema, 1991]. In Tasmania there are plans to commence a research program specifically to examine the specialist leatherwood resource. The Forest and Forest Industry Strategy of September 1990 recommended that funds be allocated to identify the future leatherwood resource, existing and future access to the resource and management techniques to maintain leatherwood honey supplies from State Forests.

Potential for on-farm production

Farmers can possibly derive incidental income from small scale honey production; returns are likely to be higher as the high transportation costs of the commercial producers will not be incurred. Nectar producing trees and shrubs can easily be incorporated into farm plantings for shade, shelter or erosion control. It is advantageous to plant trees and shrubs which flower at times when there are commonly nectar shortages [Lansdown and Reid, 1991]. Clemson recommends the following tree species for N.S.W. farm plantings because of their value to bees, ornamental and other attributes:

E. albens	White Box	inland and tablelands
E. camaldulensis	River Red Gum	inland creeks and watercourses
E. cladocalyx	Sugar Gum	inland
E. citrodoria	Lemon Scented Gum	inland
E. fibrosia	Broad Leaved Ironbark	coast and inland-clay soil
E. ficfolia	Red Flowering Gum	coast and inland
E. grandis	Flooded Gum	coast
E. leucoxylon	Yellow Gum subsp.	inland
E. longifolia	Woollybut	coast
E. maculata	Spotted Gum	coast
E. mannifera subsp. *maculosa*	Brittle Gum	coast and tablelands
E. melliodora	Yellow Box	inland
E. paniculata	Grey Ironbark	coast
E. pauciflora	Snow Gum	high altitudes-extreme cold
E. rossii	Inland Scribbly Gum	tablelands
E. saligna	Sydney Blue Gum	coast
E. sclerophylla	Scribbly Gum	coast-sandy soil
E. sideroxylon	Mugga	coast and inland-clay soil
E.tereticornis	Forest Red Gum	coast
E. torquata	Coral Gum	inland-far west
E. viminalis	Ribbon Gum	tablelands-cold conditions

Cut flowers and foliage

The cut flower industry in Australia has traditionally grown exotic species for sale, but recently there has been more emphasis on natives within the industry. Until recently the bulk of native cut flowers were collected from the bush, but the emphasis is now on commercial culture of these plants [Elliot and Jones, 1980]. In some states, particularly South Australia and Western Australia, the cultivation of native species for the export market is well advanced, whereas in other states, the flower and foliage industry is largely supplied from bush pickings or semi-managed stands.

Production, value and markets

The export of cut flowers and foliage from Australia is a significant industry, with export earnings expected to reach $50 million in 1990/91 [Windle et al., 1990]. Japan and the United States account for 65% of Australia's cut flower exports, with Holland another 14.8% [Minnis, 1990]. Western Australia produces over half of Australia's foliage and cut flowers [Pegrum, 1989]. Export earnings from W.A. have grown from $1.3 million in 1980–81 to $7.7 million in 1988–89 [Pegrum and Webb, 1990; Batt, 1990]. South Australia, another major producer, is estimated to have exported $400,000 worth of flowers in 1988/89 [Windle et al., 1990].

In Queensland, native species are being increasingly incorporated into the nursery and cut flower trade. For example, total Queensland production of wax flowers in the 1991 season is estimated at 174,000 bunches worth an estimated $400,000 at the farm gate. In excess of 60,000 bunches were exported, mainly to North America [C. Carson, pers. comm.]. In addition, there is also a substantial trade in material harvested from the wild. Approximately $25,000 was paid in royalties to the Queensland Forest Service for foliage, wildflowers and epiphytes harvested from state forests over the first 9 months of the 1990/91 financial year. At least one commercial harvester exports material to the United States and Japan while a number of smaller collectors supply the domestic market [R. Henderson, pers. comm.].

Plant species and geographic location

Cut flowers and foliage are produced from a range of native and exotic species. In Western Australia the major species harvested include banksias, boronias, dryandras, stirlingias, verticordias, waxes, kangaroo paws and numerous foliage lines such as emu bush and smoke bush [Pegrum and Webb, 1990]. The dramatic increase in production of wildflowers is partly due to the rapid expansion of privately owned plantations, particularly in the shire of Coorow (300 km north of Perth). Over 50 species of native plants are cultivated on over 800 ha of plantation accounting for some 45% of wildflower exports from W.A. [Scheltema, 1991]. The main species grown include kangaroo paws (*Anigozanthos pulcherrimus*), Geraldton Wax cultivars (*Chamelaucium unicinatum*) and banksias (*Banksia hookerianna, B.*

coccinia and *B. prionotes*) [Pegrum and Webb, 1990]. Non-cultivated wildflowers come from either natural bush stands or from managed natural stands predominantly throughout the higher rainfall areas of south west W.A.

The most important regions in South Australia for the cultivation of native flowers are the Fleurieu Peninsula, the Riverland and the south-east. Although bush picking of the Hills Daisy (*Ixodia achilleoides*) is common, most commercial cut flowers are grown under irrigation. Many of these plantations were established in the 1980s, with over 100,000 plants in the Murray Bridge/Riverland area and some 56 ha planted exclusively with banksias [Bulman et al., 1991].

In Tasmania there is a small floricultural trade of mainly foliage lines. Local ad-hoc collections of foliage from several tree species including myrtle (*Nothofagus cunninghanii*), deciduous beech (*N. gunnii*) and juvenile foliage from snow gum (*E. coccifera*), spinning gum (*E. perriniana*) and risdon peppermint (*E. risdonii*), supplemented by imports from the Australian mainland, are adequate to supply local demand. *Thryptomene micrantha* has been successfully grown as a cut flower in Tasmania for many years [Elliot and Jones, 1980] and the Tasmanian Department of Primary Industry is now investigating the potential of the Tasmanian Waratah (*Telopea truncata*) as a commercial flower for the export trade [Castley, 1991].

The Queensland cultivated wildflower industry can be categorised into two broad types, the first based on species native to Western Australia and the second based on those native to Queensland. Western Australian species have been dominant to date and there are significant plantations in southern Queensland, particularly of wax flowers (primarily *Chamaelaucium uncinatum*). However, there is now increasing emphasis on identifying and developing a range of Queensland flower species for example the rice flower (*Helichrysum diosmifolium*) for cultivation [C. Carson, pers. comm.].

In N.S.W. there is very little commercial growing of wildflowers. Most wildflowers and foliage are bush picked. There are however some popular flower species in cultivation including christmas bells (*Blandfordia* spp.), boronias (*Boronia* spp.), waratahs (*Telopea* spp.), flannel flowers (*Actinotus helianthi*), wax flowers (*Eriostemon* spp. and *Crowea* spp.) and banksia species. Foliage is harvested from a range of ferns, juvenile and new foliage of eucalypt species, leptospermums and the Gymea lily (*Doryanthes excelsa)*.

In Victoria, the Dandenong Ranges, just outside Melbourne have been a focus of cut flower growing. *Boronia megastigma*, the brown boronia native to WA, and the waratah (*T. specciosissima*) are very popular in the Dandenongs where they have been farmed for many years. *Thryptomene calycina* was also widely grown in the sand belt to the south of Melbourne in areas such as Sandringham and Brighton [Elliot and Jones, 1980]. Foliage from juvenile eucalypts, cypress, Yakka spikes (*Xanthorrhea* spp.) and Koala fern (*Restio tetraphyall*) is also collected on an ad hoc basis for floral arrangements under licence from state forests [Lansdown and Reid, 1991].

Problems and research
Problems facing the wildflower industry in Western Australia include the lack of information on: species and quantities collected by bush pickers; appropriate agronomic practices for wildflower cultivation; and appropriate techniques for bush management and marketing. The quality and supply of wildflowers for export and the spread of the root fungus *Phytophthora* are also problems for the Western Australian wildflower industry. The W.A. Department of Agriculture is currently undertaking research in these areas — including basic agronomy of wildflower production, insect problems, plant breeding and post harvest treatments of waxes, myrtles and kangaroo paws. Future research is likely to focus on increasing product quality and determining marketing strategies. However, further research is required on market strategies and promotion, fluctuations and seasonality of supply, and the sustainability of bush picking.

Research into the cut flower industry in South Australia has been carried out by a number of organisations. The WAITE Agricultural Research Institute is currently studying the development of *Helichrysum spp., Helipterum* spp. and *Anigozanthos* spp. and the storage, transport and longevity of cut flowers. The Department of Agriculture is studying the development of the Sturt Desert Pea, while the Black Hill Flora Centre is researching selection, propagation and cultivation of the Hills Daisy. The Woods and Forests Department have been researching appropriate herbicides for broadleaved weed control in cut flower crops [Bulman et al., 1991] and are currently running a project to evaluate a wide range of dripper irrigated native plants, which were established in 1987 in the Murray Bridge region. See Bulman et al. [1991] for species lists and performance.

In Tasmania the size of the local market is a problem which limits the growth of local industry. The Tasmanian Department of Primary Industry has therefore established trial plantings of the Tasmanian Waratah in the hope of developing it as a commercial cut flower for the export trade However more market research, focussed on exports of unique Tasmanian flora lines, is needed to expand the industry. Additional research requirements include insect pest and disease control for native species and information on post-harvest handling of cut flowers [Castley, 1991].

In Queensland the current emphasis is on the introduction into cultivation of wildflower species endemic to Queensland. The Queensland Department of Primary Industries Horticultural Research Station at Redlands is concentrating on selection, breeding and development of a broad range of species as well as studies in nursery techniques and post harvest handling. In some cases there are specific problems which preclude cultivation of some commercially significant bush-harvested species. For example, little is known about the propagation of koala fern (*Caustis blakei*) which is the focus of project by the University of Queensland Gatton College Campus [C. Carson, pers comm.].

In N.S.W., the N.S.W. Dept. Agriculture Horticultural Research Station, Gosford are hybridizing and developing propagation techniques for boronias, kangaroo paws, waxes and christmas bells for commercial plantings. P. Goodwen of the University of Sydney has developed a range of waratah clones which vary in colour from white to deep red. Also in progress are investigations of the flowering requirements for these species and the hybridization of paper daisies (*Helichrysum* spp.) by a commercial grower Jeremy Smith [J. Murison, pers. comm.].

Potential for on-farm production

Under appropriate conditions, cut flower growing can be a successful venture either as a crop in the conventional sense or as a component of other farm plantings (see Bulman et al. [1991] for examples of species worth considering for inclusion in stock, crop and land protection projects in South Australia). In more temperate regions, tree species with attractive foliage could also be incorporated into shelterbelts, providing additional farm income. See Elliot and Jones [1980, pp. 41—43] for a list of Australian species suitable for cut flowers and for those already grown commercially, and Elliot and Jones [1980, pp. 43—44] for species prized by florists for their ornamental foliage.

Seed collection

Production, value and markets

There are some 44 commercial seed collectors and distributors of Australian native tree and shrub seed. Most are private seed suppliers but there are also four government suppliers: The Australian Tree Seed Centre, Queensland Forest Service, Tasmanian Forestry Commission and the Western Australian Department of Conservation and Land Management [Australian Tree Seed Centre]. Little information is available regarding the quantity and value of distributed seed, however the Australian Tree Seed Centre estimate that each year they send 500 kg of seed from almost 1,000 species to over 100 countries [Australian Tree Seed Centre, undated]. Australia's commercial seed export industry is valued at over $8 million annually. Western Australian seed exports are relatively small, but have increased in value from $43,000 in 1985/6 to $213,000 in 1986/7 [Scheltema, 1991]. The majority of seeds are exported to European seed companies.

Plant species and geographic location

The Australian Tree Seed Centre concentrates on the supply of authenticated seed from eucalypts, acacias, casuarinas and other Australian native trees from most regions of Australia for research and tree improvement programs.

Tasmania has several professional seed merchants and many volunteers collecting seed from private property. The seeds of naturally occuring commercial timber species such as *E. regnans, E. globulus, E. obliqua, E. delegatensis, E. viminalis, E. amygdalina* and *Acacia melanoxylon* are in high demand for plantation programs — locally, on the mainland and overseas. However, although seed from these species attracts high prices, it is often difficult to obtain, produces a low proportion of viable seed and provenance information from natural stands is often unknown.

In Western Australia seeds of native wildflowers and trees are collected throughout the state by several small producers, mainly for revegetation projects run by local councils and mining companies, but also, in smaller quantities for domestic consumption and export.

Research

Little research on the native seed industry has been undertaken by private operators or State Government Departments, apart from a few trials on the production of wildflower and forest tree seed. The Australian Tree Seed Centre is involved in research on genetic variation and breeding systems for eucalypts, acacias, grevilleas and casuarinas. They also collaborate in research on eucalypt taxonomy, breeding strategies, quantitative genetics, isozyme studies and investigations of a range of physiological attributes such as frost resistance, salt tolerance, nutritional requirements and water relations. The Australian Tree Seed Centre supports forestry programs of the Australian Council for International Agricultural Research (ACIAR) which aim to identify and assist in the introduction and domestication of Australian tree and shrub species to meet fuelwood, agroforestry and industrial needs of developing countries [Midgely, 1988].

Potential for on-farm production

Most seed orchards are grown in much the same manner as fruit tree orchards, but there is no reason why trees grown for commercial production of seed cannot be used in an agroforestry design — that is incorporated into traditional farming operations. Further, the economic viability of any such operation is more likely if the trees are able to provide multiple uses and benefits [J. Weatherstone, pers. comm.].

Generally returns are not high, but are very dependant on species. J. Weatherstone (pers. comm.) estimates that to pick, dry, clean and store one kilogram of acacia seed requires two to three hours; the wholesale prices are generally about \$30—\$35/kg, with retail prices at \$60—\$70/kg. Eucalypts may on average take two to five hours to pick, dry, clean and store one kilogram of seed, but with much greater variability in labour time, seed quality and economic value. Retail prices start at about \$100/ kg and can be as much as \$600/kg for some species [J. Weatherstone, pers. comm.].

Essential oils

Essential oils are volatile oils obtained by the distillation of plant material. The oils may be present in any part of the plant and are used in a wide variety of products, e.g. soaps, perfumes, pharmaceuticals, solvents, cleaning agents and flavourings in foodstuffs [Cribb and Cribb, 1981].

Tea-tree Oil

Tea-tree oil was originally recognised as an effective antiseptic against a range of bacterial and fungal infections in the 1920s, but was largely displaced by synthetic antiseptics in the 1940s. Demand for the oil increased in the 1980s with the move toward natural foods and medicines. This led to the establishment of plantations on the north coast, and stimulated research in the industry [Colton and Murtagh, 1990].

Production, value and markets

The tea-tree oil industry has been very successful and is rapidly expanding: in 1987 it was valued at $340,000; by 1990 New South Wales tea-tree oil production had a value of between $4.5 and $5 million [Colton and Murtagh, 1990; Hawkins, 1991]; total production in 1990 amounted to 100 tonnes, an increase of nearly 50% from the 55 tonnes produced in 1989 [Hawkins, 1991]. The increase in tea-tree oil production can be largely attributed to the development of many new plantations — by 1990 there were 820 ha of *M. alternifolia* plantation which produced 70% of N.S.W.'s tea-tree oil.

Plant species and geographic location

Tea tree oil is commercially produced from *Melaleuca alternifolia* in the northern rivers region of the north coast of New South Wales. There are however other species which may be equally suitable for oil production including: *M. dissitiflora* from arid and semi arid regions of Australia [Brophy et al., 1989]. *Astromyrtus symphyocarpa* (syn. *M. symphyocarpa*), *M. bracteata, M. leucadendra, M. linariifolia* and *Leptospermum flavescens*) from mainly tropical regions are also considered to have commercial potential and warrant further study [Brophy et al., 1989].

Problems and research

To date research into tea-tree oil production has focussed on oil chemistry and agronomic aspects of the crop (for plantation development). Current research programs are examining the pattern of growth and oil formation, the effect of water stress on production, optimal time and frequency of harvest and the efficiency of distillation [Robinson, 1988]. Other areas of research include:

- Bioassays of oil to determine its efficacy against a range of microorganisms; R. L. Williams at Macquarie University.
- Relationship between insect infestations and oil constituents; N. Treverrow/C. Maddox, N.S.W. Dept. Agriculture, Wollongbar Agricultural Institute.

- Techniques for propagation of high yielding clones; J. Wish at University of New England (UNE).
- Genetic variation in tea-tree; P. Butcher, Forestry Dept., Australian National University (ANU).
- Development of a breeding program for the industry; J. Doran, CSIRO Division of Forestry and Forest Products Research, ACT.
- Economic returns from tea-tree oil production; T. Reilly, N.S.W. Dept. Agriculture, Wollongbar Agricultural Institute; D. Etherington, UNE, Armidale.

Potential for on-farm production
Tea-tree oil production is already estimated to have contributed to the conservation of over 1 million trees as these trees have previously been extensively cleared [Cribb and Cribb, 1981]. Combined with the 30 million tea-trees already established in plantations, continuing plantation development in the northern rivers region represents one option for farmers to diversify incomes and to have a positive effect on the environment by the re-establishment of native vegetation [Pearce, 1991]. However, tea-tree grown under plantation conditions is intolerant of weed competition, and when compared with other agroforestry crops, is labour intensive. Returns have fluctuated considerably over recent years due to changing supply and demand. Currently the farm gate price ranges from \$45—\$70/kg with average production estimated at 192 kg/ha/yr. However, establishment and maintenance costs are high (in the order of \$17,000/ha and \$2,800/ha/yr respectively) [Murtagh and Etherington, 1990]. *Melaleuca alternifolia* requires acid soils and a frost free environment, and there is an 15—18 month lead time from establishment to first harvest [Hawkins, 1991].

Eucalyptus oil

Production, value and markets
Since the 1950s there has been a considerable decline in Eucalyptus oil production in Australia due to competition from China, Spain, Portugal, Chile, Swaziland and South Africa. Australia in fact now imports lower grades of Eucalyptus oil, used mainly in cleaning agents, although still exports higher grade oil (with a higher cineole content) used in the pharmaceutical industry [Boland et al., 1991]. Current world production is between 2,000 to 3,000 tonnes of which Australia produces some 3%. The world's largest Eucalyptus oil producer is China which, like most of the major producers, has established eucalypt plantations with the primary objective of wood production, but coincidentally the species planted also produce good yields of oil. Currently Eucalyptus oil production from New South Wales (Australia's biggest producer) is valued at between \$1.5 and \$2 million per year [Hawkins, 1991].

Plant species and geographic location
Of the 800 species of Eucalyptus only 20 are exploited commercially and 12 are of economic importance for oil production [Bulman et al., 1991; Boland et al., 1991]. The industry is predominantly based on the production of cineole from Blue Mallee (*E. polybractea*) which occurs in disjunct populations in low rainfall areas near Bendigo (Vic.) and the West Wyalong area of central western N.S.W. [R. James, pers. comm.]. A private company, G.R. Davis Pty. Ltd. has established a Blue Mallee plantation at West Wyalong, in addition to enrichment planting of natural blue mallee stands in the area.

Problems and research
Research into Eucalyptus oils is currently underway at several centres including the N.S.W. Department of Agriculture at Condobolin, the Department of Forestry at the Australian National University, the University of N.S.W., Murdoch University, The University of Tasmania and C.S.I.R.O. Divisions of Plant Industry and Forest and Forest Products Research in the A.C.T.. The N.S.W. Department of Agriculture and CSIRO Plant Industry have established trial plantations of *E. polybractea* and *E. kochii* at Condobolin, to assess the potential of these species as major oil producing crops. *Eucalyptus kochii* is a mallee or small tree which grows in the wheat growing areas of south-west Western Australia, within the 280—380 mm rainfall isohyets. Recent trials in W.A. have shown that the oil yield for *E. kochii* ranges from 1.2%—3.7% of fresh leaf weight [Brooker et al., 1988], compared to that of *E. polybractea* at 0.7%—5.0% [Lassak, 1988]. The University of N.S.W. is examining the chemistry of Eucalyptus oils, while the C.S.I.R.O. Division of Plant Industry is studying chemotaxonomic aspects and comparative production between species.

Although some research into the production of Eucalyptus oil in both South and Western Australia has been undertaken, there is currently no commercial production in either state due to the high costs of production and small returns [Western Australian Department of Agriculture, 1989; Bulman et al., 1991]. Recently, however, a number of trials have been set up to look at prospects for combining oil production with land conservation. A number of eucalypt species (*E. angustissima, E. calycogona, E. dives, E. horistes, E. loxophleba, E. spathulata* and subspecies of *E. kochii*) are currently being trialled for oil and cineole content and water usage at the Wongan Hills research station by the Western Australian Department of Agriculture and Murdoch University — with some encouraging results [see Scheltema, 1991].

During the 1980s the University of Tasmania and the private company Felton, Grimwade and Bickford Pty. Ltd. established a number of trials to determine the oil content of *E. radiata var. australiana.* Although the trees grew well under Tasmanian conditions, oil content yields were disappointing [see Castley, 1991].

The focus of joint research programs at the A.N.U. Dept. of Forestry and

CSIRO Division of Forestry and Forest Products Research is on plant breeding and improvement, although there is some investigation of the biological functions of the oil in leaves and the response of oil yield to environmental conditions. Three species are currently being examined: *E. polybractea* and *E. kochii* (for production within Australia) and *E. camaldulensis* (for oil production as a byproduct of fuelwood plantations established in many developing countries in the wet/dry tropics) [see Doran and Matheson, in press]. The aim is to improve the quantity of foliage, the oil concentration in the foliage (particularly cineole), the percent of biomass as foliage and the speed of growth.

Potential for on-farm production

Methods of oil production vary from species to species and as a result of differing management regimes. In semi-arid N.S.W. where the bulk of Australian Eucalyptus oil is produced, the blue mallee is well suited to commercial production because coppice regrowth can be harvested every two years. In contrast Eucalyptus oil production in China (the world's largest producer) and elsewhere, is largely a byproduct of logging residues with cut branches being processed in small family or community owned stills [P. Evans, pers. comm.].

In Australia there may be potential for similar small scale production of Eucalyptus oil, either as a byproduct of logging privately owned native forest or as an additional product from agroforestry plantings. With approximately 200 ha of suitable species available as a resource, a small still could be set up for a few thousand dollars [Abbott, 1986] either by a single landowner or as a co-operative venture between a number of landowners. Trials have been established to examine the viability of similar ventures in Western Australia [Barton and O'Reilly, 1986].

Other essential oils

The essential oils industry in Tasmania has concentrated on crops such as fennel, peppermint, boronia, parsley, blackcurrant, spearmint, carraway and dill [Castley, 1991], although a valuable oil used as a preservative and for medicinal purposes was formerly produced from sawdust from the milling of Huon Pine [Cribb and Cribb, 1981]. More recently, attention has been focussed on the oils of other native species including: cider gum (*E. gunnii*) which yields a sugary syrup which could be used as an alternative to maple syrup; Sassafras (*Atherosperma moschatum*) which produces a culinary essence; and native pepper (*Tasmania lanceolata*) which produces a peppery spice [Harris, 1990].

The Western Australian essential oils industry is small, with the only commercially viable oil produced from the Brown Boronia (*Boronia megastigma*) for perfume. The Brown Boronia is the only native Western Australian flower that has been commercially exploited for its perfume and is sold both locally and overseas [Cribb and Cribb, 1981].

There is also some interest from the flavour and fragrance industries in the production of E-methyl cinnamate from Stringybark peppermint (*E. olida*). Initial studies have shown that returns from small scale commercial distillation can be as high as $400/kg and that the species is well adapted to coppicing and plantation situations. Currently the industry is centred around natural stands on freehold land in the New England region of N.S.W. where the species occurs naturally [Robinson, 1988].

Medicinal plants

Queensland rainforests are a rich potential source of medicinal plants in Australia. There is an extensive range of chemicals primarily alkaloids for pharmaceuticals and medicines which can be derived from rainforest plants and form the basis of an established industry in Queensland [Ryan et al., 1991].

Duboisia

Production, value and markets

The Duboisia industry is largely controlled by European drug companies (Boehringer-Ingelheim operates the largest plantation) who exported 500 tonnes of dried leaf in 1988/89, conservatively estimated to be worth in excess of A$10 million when processed [Ryan et al., 1991].

Plant species and geographic location

The most important medicinal tree grown commercially in Queensland is a hybrid between *Duboisia leichhardtii* (a native of drier regions of south central Queensland and the far western plains of N.S.W. [Lassak and McCarthy, 1983]) and *D. myoporoides* (which grows at the edge of rainforests from Sydney to Cape York Peninsula [Lassak and McCarthy, 1983]) [Lake, 1990]. Duboisia leaves contain tropane alkaloid drugs scoplamine and hyoscyamine from which atropine is manufactured [Lake, 1990].

Problems and research

Extensive investigations of Queensland rainforest flora for constituents of value in medicinal drugs have been carried out by the Australian Phytochemical Survey since 1940. Although some 2,000 papers on the chemistry of medicinal plants have been published [Ryan et al., 1991], the project has failed to encourage further commercial production of medicinal plants [Lassak and McCarthy, 1983]. But the major reason for the neglect of our medicinal plant resources is attributed to the absence of a research-based pharmaceutical industry in Australia and the fact that only a joint operation of cultivation and processing is seen as a viable enterprise [Warren, 1989].

Potential for on-farm production

Duboisia plantations have been established in the Burnett area in Queensland. Duboisia is a fast growing crop which requires neither irrigation nor fertiliser, and under optimum conditions the leaves can be harvested after the first year [Lake, 1990]. These are then dried and mostly sent overseas for further processing [Lake, 1990].

Other potential medicinal plants

Another potential commercial medicinal plant is *Solanum laciniatum*, which also grows in Queensland rainforests [Warren, 1989]. This plant produces solasodine, primarily in the unripe fruit, which is used as a chemical precursor for the synthesis of steroidal contraceptives [Lassak and McCarthy, 1983]. Attempts have been made to establish plantations and set up manufacturing facilities in Australia, but without the support of government funding the projects were not undertaken [Warren, 1989]. Both Hungary and Russia have now established extensive plantations of *S. laciniatum/aviculare*, leaving Australia with no solasodine industry at all [Warren, 1989].

Nut trees

Macadamia nuts

Production, value and markets

In 1990 New South Wales growers produced 6,100 tonnes of macadamia nuts with a $15 million farm gate value [Hawkins, 1991] while Queensland growers produced 2,254 tonnes (1988/89) with a gross value of $3,500/tonne [Ryan et al., 1991]. Returns to growers in N.S.W. are in the range of $3—$4/kg.

Plant species and geographic location

The macadamia (*Macadamia integrifolia* and *M. tetraphylla*) is currently the only Australian native tree developed as a commercial food crop. Macadamias are rainforest trees, indigenous to the coastal areas of northern N.S.W. and south eastern Queensland, between latitudes 25 and 32 South [Hawkins, 1991; Ryan et al., 1991]. Although there are increasing areas of macadamias under cultivation (e.g. in N.S.W. plantation area increased from 1,500 ha in 1980 to 4,400 ha in 1991), there are far more extensive plantations in Hawaii than in Australia [Hawkins, 1991; Cribb and Cribb, 1974]. There are also extensive plantings in South Africa, Rhodesia, Kenya, California and Central America [Hawkins, 1991].

Potential for on-farm production

Macadamias require frost free, sheltered and well drained sites for successful establishment in plantations. The minimum economic size for a viable orchard is 10 to 20 ha, depending on land prices. Grafted *M. integrifolia* crop lightly after 4 years (approx 1kg/tree) and increase production with age, levelling out at approximately 30 kg/tree from year 20 onwards [Hawkins, 1991].

They require rain in early spring and summer for good fruit set, but irrigation can be successful in commercial orchards [Hawkins, 1991].

Walnuts

Production, value and markets

World production of walnuts is around 400,000 tonnes, the bulk of which is produced in the U.S.A. [Allen, 1986]. Of this, Australia produces only 120—150 tonnes, mostly from north-east Victoria [Allen, 1986]. There is a large market for walnuts worldwide and Australian growers are confident that their nuts will sell well in Australia, displacing annual imports of around 2,750 tonnes (1986/87 figures). The trees could also provide income from the seeds, the shells and the high quality timber which currently sells for approximately $\$3,000/m^3$ [Lansdown and Reid, 1991].

Plant species and geographic location

Black walnut (*Juglens nigra*), a native of eastern USA, is commonly used as a rootstock for the English walnut (*J. regia*), a native of the region from Eastern Europe through to the Middle East [Allen, 1986]. These cultivars are suitable for planting in deep, well drained, fertile soils with ample water [Allen, 1986]. Most of Australia's commercial walnuts are currently grown in the river valleys of north-east Victoria and East Gippsland with small amounts grown in other regions of Victoria, including the Dandenongs and the Mallee region around Mildura. In other states, walnuts are grown in the Blue Mountains, west of Sydney, in the Riverland and Adelaide Hills region of South Australia, in south-west Western Australia and in Tasmania [Allen, 1986]. Walnut plantations for both nuts and timbers are currently being actively promoted as agroforestry crops in Tasmania and Victoria.

Problems and research

Melbourne University students are currently monitoring agroforestry plantings of walnuts on Walnut Island. Initially vegetables were grown between the rows of Walnut trees, but now Hereford cows and calves graze between the rows at similar stocking rates to those on surrounding farms. The students are monitoring grass quality, weed control and fertilizer requirements. The Forest Research Institute of New Zealand is studying thinning programmes for walnut tree agroforestry programmes [Lansdown and Reid, 1991].

Potential for on-farm production

In Tasmania a private company, Vecon Pty. Ltd., has put forward a walnut sharefarming proposal with cultivars developed by Vecon. Capital costs are expected to range from $3,500—$4,500/ha in the first year with ongoing maintenance costs of $300—$1,000/ha/yr. Presently project costs are being matched dollar for dollar by the Horticultural Research and Development Corporation and the Tasmanian Development Authority is making available development loans under its Agricultural Plan Scheme [Castley, 1991].

Chestnuts

Production, value and markets

World production of chestnuts is around 400,000 tonnes, mostly produced by Japan and Italy. Australia produces only 120—150 tonnes, but production is expected to increase to 3,500 tonnes by the year 2000 as a result of new plantings [Allen, 1986]. Chestnut trees, like walnut trees can produce not only saleable nuts (currently selling between $7—$12 kg retail) but also a high quality timber.

Plant species and geographic location

In Australia the most common species of chestnut is the Spanish chestnut *Castanea sativa*. Chestnuts grow well in cool temperate areas and are suited to fertile and well watered sites. Main production areas in Australia are in Victoria: around Bright and Beechworth in the north-east; in the Dandenongs; in the Otway Ranges; and around Ballarat. Smaller quantities are produced in the Batlow and Blue Mountains areas of N.S.W., in the Adelaide Hills of South Australia and south-west Western Australia [Allen, 1986].

Potential for on-farm production

Chestnuts are currently being promoted as a multipurpose agroforestry crop in Victoria. They are ideally suited to agroforestry regimes because of the wide spacings (56 to 70 trees per ha). Intercropping with vegetable or berry crops in the early years has been shown to be profitable and provides a cash flow while waiting for the trees to start producing [Allen, 1986]. Mature trees can also be harvested for timber, but lateral branches need to be pruned regularly to produce a clean bole suitable for valuable, high quality timber.

Problems and research

Although the chestnut in Australia is relatively free of insect pests, chestnuts are susceptible to fungal attack commonly by *Phytopthora* and *Armillaria.* The most serious disease is collar rot *Phytopthora cinnamomii*, which can cause the death of trees, particularly in poorly drained soils.Unfortunately collar rot in chestnuts is difficult to control. Although glasshouse experiments with two fungicides Ridomil and Alliette have successfully controlled collar rot, field trials have been unsuccessful [Allen, 1986].

Pistacio nuts

Production, value and markets.

Iran, Turkey, Afghanistan and Greece are the world's major producers and exporters of pistacio nuts. The pistacio was virtually unknown in Australia until the late 1960s when experimental plantations were set up by C.S.I.R.O. At present in Australia there are 4 large pistacio plantations and many smaller ones. The large plantations are located at Telopea Downs, Victoria (100 ha), Robinvale, Victoria (120 ha), Pinaroo, SA (200 ha) and at Kyalite in N.S.W. [Lansdown and Reid, 1991]. These large plantations could potentially produce up to 75% of Australian demand, which is currently met by American imports [Lansdown and Reid, 1991]. The trees produce nuts from a young age, but it takes about seven years until they produce between 5 and 10 kg of dried nuts in the shell which sell for $15 to $20/kg [Lansdown and Reid, 1991].

Plant species and geographic location
Pistacios are natives of western Asia and Asia minor. They grow naturally in climates where summers are long, hot and dry and the winters moderately cold [Allen, 1986]. In Australia the cultivated pistacio *Pistacia vera* L. is grafted onto a rootstock of *P. chinensis*, *P. atlantica* or *P. terebinthus*. As these hardy trees can tolerate low rainfall and saline conditions pistacio nut production is seen as a potentially valuable enterprise in semi-arid Australia [Lansdown and Reid, 1991]. Even under drought conditions the trees will continue to produce nuts, although in reduced quantities.

Problems and research
The C.S.I.R.O. established a research plot of 1 ha of pistacios at Merbein (Victoria) in 1968, and has more recently established another 1 ha plot in which 600 hybrid seedlings have been grafted onto rootstock [Lansdown and Reid, 1991]. The C.S.I.R.O. is examining the tolerance to varying concentrations of salt in water, propagation techniques, variety trials and breeding studies to improve nut quality [Lansdown and Reid, 1991]. Further research on propagation methods is still seen to be important as the pistacio is not easy to propagate.

Potential for on-farm production
A recent survey by the Pistacio Growers Association showed that the income from pistacios could be similar to that of walnuts or chestnuts. They estimate that with an interest rate of 17%, it would take 7 years from planting until the first return, and that in 13 years the project would break-even [Lansdown and Reid, 1991].

Sandalwood

Production, value and markets
The sandalwood industry was one of Western Australia's earliest industries and at one stage was the state's largest export earner [Scheltema, 1991]. Today sandalwood sells for $3,500—$10,000/tonne [Bulman et al., 1991] with the industry valued at $10 million annually [Scheltema, 1991]. Over 2,000 tonnes are exported to Taiwan, Hong Kong, Singapore, Thailand and Malaysia each year [Keally, 1989] where it is predominantly used to manufacture Joss sticks [Scheltema, 1991]. In Western Australia, sandalwood production is controlled by export quotas which are set according to availability, market and conservation needs [Scheltema, 1991]. South Australian exploitation has dwindled since the 1940s and sandalwood is now a protected species [Bulman et al., 1991].

Species and geographic location

Sandalwood (*Santalum specalum*) is a small, non-obligate, parasitic tree which grows in the arid and semi-arid regions of South Australia and Western Australia. There is a long history of sandalwood exploitation from both these regions as there is a good market for the valuable scented timber as well as for the oil which is used for perfumery [Cribb and Cribb, 1981]. Because the timber is valuable, the oil is extracted primarily from the roots and butts of the tree [Cribb and Cribb, 1981], which are then chipped, powdered and bagged before export [Scheltema, 1991].

Problems and research

To date, the industry has exclusively exploited natural stands of sandalwood, but due to concern about the dwindling resource, efforts are now being made to establish plantations of sandalwood [Scheltema, 1991]. Research has been carried out by the Woods and Forests Department in South Australia and by the former Western Australian Forests Department [Scheltema, 1991; Bulman et al., 1991]. From 1895 to 1981 research focussed on the resource and the germination and regeneration of sandalwood [Loneragan, 1990]. In 1988 the Sandalwood Conservation and Regeneration Project was established by the Department of Conservation and Land Management and private industry to improve the conservation status of sandalwood. As part of the project, reserves are being established on former pastoral leases and trial plots of sandalwood plantations have been established near Geraldton and Narrogin (WA), [Scheltema, 1991]. Indian sandalwood (*S. album*), which has a faster growth rate and higher oil content is also being trialled on a 20 ha plantation in an old irrigation area near Kununurra [Scheltema, 1991].

Due to low germination rates, slow growth rates and the difficulty involved in propagation and establishment (because of the parasitic nature of the plant) [Scheltema, 1991; Bulman et al., 1991] a number of institutions are researching methods of re-establishing sandalwood. Current research at Curtin University is focussed on the selection of a primary host at the pot stage for Indian sandalwood, while at Murdoch University examining methods of tissue culture propagation of Indian sandalwood are being examined [Scheltema, 1991]. As the Santalum genus has edible fruits (quandongs) Curtin University is looking at the potential of fruit production of cultivated *S. spicatum* in the Avon Valley and Narrogin areas [Scheltema, 1991].

Potential for on-farm production.

At present the industry relies on natural stands of sandalwood. Although it is possible to develop sandalwood plantations, the plant's parasitic nature means propagation and establishment are difficult and expensive. According to Bulman [1991] establishment costs for 500 plants/ha would amount to approximately $1,500/ha. Assuming a growth rate of 0.25 tonne/ha/year, a rotation length of 50 years, no fencing land or maintenance costs and a discount rate of 7%, the value of standing (unharvested) sandalwood would need to be more than $3,500/tonne to be viable. At present, although sandalwood sells for between $3,500 and $10,000/tonne, the bulk of this goes on extraction and transportation costs leaving significantly less for the grower. Proximity to markets may alone determine the viability of a sandalwood plantation, although inclusion in multipurpose plantings may be a satisfactory alternative for small scale production to obtain incidental income.

Potentially viable industries

Tannins

Tanning greatly increases the durability, water resistance and flexibility of animal hides, which if not treated would decay rapidly [Cribb and Cribb 1981]. Until recent times, most tanning processes have used tannins of vegetable origin. Tannins are also present in a wide range of flowering plants and conifers, with highest concentrations usually found in the bark and insect galls [Cribb and Cribb 1981].

Wattle bark is one of the most important tannin sources in the world. Last century tannin extraction from wattle bark was a major industry in Australia. By 1907 Australian production of all tanning bark was over 43,000 tonnes, of which 18,000 tonnes were exported. Much of this production was supplied from natural stands of *Acacia pycantha* and *A. mearnsii* in New South Wales, Victoria, Tasmania and South Australia and from *Eucalyptus astringens* in the semi-arid regions of south-west Western Australia [Cribb and Cribb, 1981; Searle, 1991]. Since the beginning of the century production of wattle bark tannin has declined dramatically. 'No single reason for the decline of the bark industry over the last century has been identified. However, the lack of investment and successful research initiatives, the unavailability of labour, competition from tannin imports and tannin substitutes, the decline of the leather industry and environmental hazards such as fire, drought, frost and insects have all contributed' [Searle, 1991 p 30]. Annual imports of tanning extracts now exceed 10,000 tonnes [Australian Bureau of Statistics March 1990—March 1991]. Of the imported tannin 4387 tonnes was wattle bark extract principally from *A. mearnsii* plantations in South Africa, valued at over $4 million [Searle, 1991; Bulman et al., 1991].

During the early part of this century commercial plantations of *A. pycnantha* and *A. mearnsii* were established in South Australia and Victoria [Bulman et al., 1991; Cribb and Cribb, 1981]. Although presently it may not be economically viable to re-establish plantations expressly for tannin production, there is a small and increasing demand for *A. mearnsii* tannin extract in adhesives used in the composite board industry. It is also possible that environmental considerations may lead to increased use of vegetable tannins in the leather industry [Searle, 1991]. The future may well lie in the integrated management of *A. mearnsii* plantations for pulp, tannin, fuelwood and reconstituted wood products [Searle, 1991] or the multiple use of plantations of other species which also provide a source of tannin. For example, bark from *Pinus radiata* is a potential major source of tannin. Investigations in the 1950s and 1970s examined the feasibility of extracting *P. radiata* tannin for general tanning and for use as an adhesive in the particle board industry [Bulman et al., 1991]. Yields of tannin were lower than those of wattle bark but still satisfactory for tanning and adhesive manufacture [Anderson and Bilkens, 1958]. One private company is currently believed to be in the process of extracting tannins from bark from *P. radiata* plantations in the south-east of South Australia [Bulman et al., 1991].

Non-indigenous tree crops

Christmas trees

The bulk of Christmas trees comes from the thinnings of commercial state and private plantations. Numerous species are used including Douglas fir, Dutch pine, Norway spruce and Monterey pine [Lansdown and Reid, 1991]. However, in Victoria two plantations have been established for specialist Christmas tree production. The first, located in the Dandenongs, has been growing Christmas trees for over 30 years. The other has about 40 ha of plantation on the Nepean Peninsula and shears radiata pine using technology developed in America to produce neat trees with dense foliage which command a premium price [Lansdown and Reid, 1991]. Alternative species have been trialled for appearance and growing time, but most are considered too slow growing for mass production [Lansdown and Reid, 1991].

Carob

Carob (*Ceratonia siliqua*) is a traditional Mediterranean crop. In Australia there is a strong and growing demand for carob syrup and beans for human consumption and as a stock feed supplement. Imports of carob have grown from 1200 tonnes in 1974 to 1500 tonnes in 1988 [Bulman et al., 1991].

Carob has been grown in Australia for over 140 years in a range of soil types and climatic conditions. The trees will grow in limestone, sands and cracking clays and will tolerate annual precipitation as low as 250 mm [Bulman et al., 1991]. It is a useful species for general shade and shelter, which grows in dry temperate to semi-arid situations, and has the added benefit of producing palatable pods for stock.

However, for commercial production of carob powder it is recommended that superior material is grafted onto root stocks and that trees grown in areas with less than 600 mm rainfall per annum be irrigated. Some research has been carried out on development of grafting techniques to improve pod quality for commercial production [see Bulman et al., 1991]. Yields for commercial production are as yet unknown, but it is speculated that a maximum of 25 tonnes/ha could be produced with grafted stock under irrigation, possibly worth $2000—$3000 per tonne when processed [Bulman et al., 1991]. Economic analysis of growing carobs commercially has been carried out by Esbenshade & Wilson in 1986; they report that 'with basin irrigation and about a four tonnes per hectare yield by the 15th year, a positive net annual return could be achieved as early as the seventh year if the pods were kibbled and sold at their highest value for carob powder, and the seeds sold for locust bean gum'. Significantly higher returns were reported for plantations under trickle irrigation.

Dates

Date palms (*Phoenix dactylifera)* are grown in arid and semi-arid regions of the middle east and north Africa for their fruit, timber and root fibres (which are used for padding in furniture and car seats) [Lansdown and Reid, 1991]. Date palms are very hardy plants that are able to withstand a temperature range of −16 °C to 52 °C (but do not fruit under 18 °C). They thrive in coastal areas where their roots are continually flushed by sea water [Lansdown and Reid, 1991]. Date palms produce an average of 30 to 40 kg of fruit annually and up to 120 kg in ideal conditions [Lansdown and Reid, 1991].

Australia imports some 4000 tonnes of dried dates annually which sell in fruit markets for $16—$18/kg [Lansdown and Reid, 1991]. A number of commercial date palm plantations have been established in Australia to supplement imports, primarily in Queensland with some in the Northern Territory [Lansdown and Reid, 1991].

A number of trial plantations have been established in Western Australia and Victoria primarily for their ability to withstand high saline water tables. The Western Australian agricultural regions are not suitable for date production as summer time temperatures are insufficient for the palms to produce fruit [Scheltema, 1991].

In Victoria a joint study between the Departments of Conservation & Environment, Agriculture & Rural Affairs and the Rural Water Commission was initiated in 1989 to investigate the effects of the date palm on the water table at a 4 ha site near Swan Hill [Lansdown and Reid, 1991]. The date palms planted originated from Saudi Arabia, Alice Springs and Egypt. While those from Saudi Arabia and Alice Springs have all but died, those from Egypt are growing well. When suitable palms are found it is hoped that further research will be done trying to find varieties which will fruit in commercial quantities [Lansdown and Reid, 1991].

Broombush

Broombush (*Melaleuca uncinata*) is used for the construction of brush fencing. Broombush naturally occurs in semi-arid regions of South Australia and the Wimmera/Mallee region of Victoria and has been harvested from natural stands in these areas for many decades.

Recently concerns have been voiced at the sustainability of broombush harvesting [Bulman et al., 1991]. For example, Newland [in press] reports that South Australian broombush harvesting has increased from 1800 20 kg bundles per week in 1979, to 2650 20 kg bundles per week in 1986. Current research projects in both Victoria and South Australia have concentrated on the establishment of Broombush plantations [Wismiewski in Lansdown and Reid, 1991; Bulman et al., 1991; Newland, in press] with the eventual aim of supplementing or substituting broombush supplies from natural stands.

John Wismiewski, from Melbourne University is studying the feasibility of establishing *Melaleuca uncinata* on marginal Mallee/Wimmera farmland, for supplementing income from traditional cropping and grazing practices in the area [Lansdown and Reid, 1991]. He is currently trialling three different provenances of Broombush, examining growth rates, developing suitable establishment techniques, determining fertilizer requirements and assessing the effects of plant spacing and weed competition [Lansdown and Reid, 1991]. Preliminary economic studies of on-farm production of broombush in South Australia show that it is not a profitable proposition unless present royalty rates (5c–\$1.00 per bundle) rise markedly [Bulman et al., 1991; Newland, in press]. Bulman et. al. [1991] compared the costs and returns of three establishment methods for broombush: non-irrigated seedlings, direct seeding and irrigated seedlings. They showed that establishing broombush plantations using direct seeding would not lose money, but would return only approximately \$8.50/ha/year; irrigated plantations established with seedlings would only be profitable under the most optimistic circumstances; non-irrigated seedlings would not be profitable.

Possibly viable industries

Craftwood

Tasmania has an internationally respected craftwood industry which is attributed to the quality timber species available. Their most important species are Huon pine (*Lagarostrobos franklinii*), King Billy Pine (*Athrotaxis selaginoides*), Blackwood (*Acacia melanoxylon*), Myrtle (*Nothofagus cunninghamii*) and Sassafras (*Atherosperma moschatum*). These species are often made available as a result of logging operations in state forests, sometimes in quantities far in excess of requirements [Castley, 1991].

Galls, or burls are becoming increasingly popular as craftwood; they are strange shaped protuberances on the leaves or stems of plants, produced by a great variety of insects and fungi, and are particularly common on *Eucalyptus* spp. and *Acacia* spp. [Wrigley and Fagg, 1979]. A major problem which is currently besetting the craftwood industry is the indiscriminate felling of quality trees purely for the purpose of obtaining burls or galls [Castley, 1991].

Little research has been carried out which is applicable to the craftwood industry, although in 1989 the University of Tasmania and the Tasmanian Forestry Commission attempted to produce the highly desirable 'Black Heart' (a light to dark brown discolouration in the central core of the trunk of trees with pale coloured heartwoods, [Hillis, 1987]) by inoculation with a variety of fungi in clear grained Sassafras timber. *Penicillium enchinlatum* proved to be the most effective of the fungi tested and further large scale inoculation trials are planned [Tasmanian Forestry Commission, 1989/90].

The craftwood industry in Western Australia is centred in the high rainfall zone in the south-west region of the state. Furniture production dominates the industry (47% of production), followed by woodturning (23%) and sculpture (22%) [Scheltema, 1991]. Although the craftwood industry is cottage based, with 85% of producers being hobbyists, the industry is valued at over $2 million per year [Scheltema, 1991]. Between 1,700 and 2,800 cubic metres of timber are used annually, collected from the forest floor or purchased from retailers. It is estimated that 80% of this timber comes from crown lands [Scheltema, 1991].

The timbers used in the craftwood industry are selected for their unusual colour, grain or texture. The most commonly used species in Western Australia is Jarrah (*E. marginata*) followed by she-oaks (*Casuarina* sp.), Blackbutt (*E. patens*), *Banksia* spp., Marri (*E. calophylla*) and yate (*E. cornata*) [Scheltema, 1991]. The final products are sold through galleries and craftshops (40%) or by private commission (40%).

Research to date by the 'Fine Wood Project' has focussed on the identification of participants, products and markets. There are plans for further research into the potential for growing fine wood timbers, the assessment of volumes of craftwood recoverable from forestry operations, health and safety, and the export market for craftwood products [Scheltema, 1991].

Charcoal

High quality charcoal is produced from many of the slow growing trees of the arid and semi-arid zones of South Australia. Species include *Eucalyptus anceps, E. dumosa, E. foecunda, E. gracilis, E. socialis, Casuarina cristata, C. leuhmanii* and *Acacia aneura* [Bulman et al., 1991]. Many other species from other regions have also been used to produce charcoal including: *E. cladocalyx, E. largiflorens, E. leucoxylon, E. obliqua* and *E. camaldulensis* which is still being used today to produce charcoal in the Barmah/Millewa forests of the Murray river basin in N.S.W. and Victoria.

During the Second World War charcoal was used extensively to produce gas as a petrol substitute [Bulman et al., 1991]. Charcoal Industries of South Australia still produce much the same quantities (30 tonnes per week in the early 1980s) in the same way but it is used for different applications [Bulman et al., 1991]. The major uses are now for smelting, cooking (char-grilling in restaurants), agricultural charcoal (animal foods, potting mixes and horticulture) and in the automotive industry [Bulman et al., 1991].

It takes between 2 and 2.5 tonnes of wood to produce 1 tonne of charcoal [Bulman et al., 1991] and with returns to growers of approximately $5/tonne for access to timber, farmers could not expect to get a profitable economic return on a plantation expressly for charcoal production. However, it may be possible to generate additional income from charcoal by incidental harvesting of fuelwood and multiple use of shelter plantings [Bulman et al., 1991].

Quandongs

As previously mentioned, *Santalum* spp. produce edible fruits. *Santalum* spp. are non-obligate parasitic trees restricted to arid and semi-arid regions [Cribb and Cribb, 1974]. The sweet quandong (*Santalum acuminatum*) has particularly large and tasty fruit [Isaacs, 1987]. When ripe their colour ranges from incipid yellow to deep crimson [A. Beal, pers. comm.] and they are 2—3 centimetres in diameter [Cribb and Cribb, 1974]. These fruits were very popular with Aborigines and were also commonly used by early European settlers to make jams, pies and jellies [Cribb and Cribb, 1974].

The C.S.I.R.O. Division of Horticulture has been conducting research into the domestication of the quandong since 1973 [Bulman et al., 1991]. The aim is to produce high yielding cultivars which bear highly palatable fruit reliably, are drought and salt tolerant and resistant to pests and diseases. The C.S.I.R.O. now has a short list of genotypes for developing large fruit, free of bitterness.

Some farmers in S.A. have successfully established orchards of quandongs, which sell for an average price of $30—$40 per kilogram, although prices of $50—$60 have been quoted [Beal, pers. comm.]. One farmer at Quorn (S.A.) has successfully irrigated a quandong orchard with saline bore water (at 4,000 ppm) [Bulman et al., 1991].

There have also been some attempts to establish quandongs on farms in the arid and semi-arid regions of Western Australia. Scheltema [1991] reports that attempts to establish orchards on farms in the sandplain country north of Geraldton have been unsuccessful. However, quandongs have been established in the Chapman Valley, south of Geraldton.

As with establishing *Santalum* spp. for craftwood, plantation establishment for quandong production provides major difficulties. Considerable research is required to develop propagation techniques, plant selection and breeding, cultivation practices, market research and promotion.

Although a number of small scale plantations have been successfully established, potential and existing growers should proceed cautiously [A. Beal, pers. comm.].

Cork oak

Cork is the bark of *Quercus suber*, the cork oak tree. The species is indigenous to countries bounding the western Mediterranean where it occurs naturally as an evergreen open woodland tree [Cooke, 1961]. The species grows well on a wide range of soils in Mediterranean climates — it is a strong competitor having extensive deep and shallow roots [J. Banks, pers. comm.]. The natural form of the cork oak has a short bole and large spreading crown. Trees of 20 metres height with boles of 4 to 5 metres in length are common, dependent on growing conditions.

Cork is composed of dead watertight cells and as such is light and buoyant. It is used for bottle and jar seals, insulation against noise, temperature and vibration, floats and for shoes [Cooke, 1961]; [J. Banks, pers. comm.].

The acorns of the cork oak tree are also used as stock feed; pigs traditionally free range under the trees during periods of acorn fall in the Mediterranean [Cooke, 1961]. The cork oak tree also makes a good ornamental tree; as a single tree the form is large and spreading, while in plantations the closed canopy can produce majestic 'radiating cathedral like avenues' [Banks et al., 1990, p. 8].

Australia produces little or no cork of its own — although the cork oak is known to grow in many diverse regions of Australia including the Hunter Valley in N.S.W., the Dandenong Mountains near Melbourne, Victoria and the Mount Lofty Ranges close to Adelaide, South Australia. Despite this, the only sizeable cork oak plantation known to the authors is the Glenloch Plantation in the Australian Capital Territory. The plantation is listed on the Register of the National Estate and on the ACT Sites of Significance Register and is no longer harvested [Banks et al., 1990].

In the financial years of 1989—90 and 1990—91 Australia imported $1.2 million worth of Cork, mainly in granulated, crushed or ground form; some was then exported after processing [Australian Bureau of Statistics, 1991]. Experience with the 4,500 trees at the Glenloch Plantation demonstrated that commercial stripping is possible. Regular harvests were made from 1948 until 1964 and again in 1979 when 10 tonnes of cork were removed [Banks et al., 1990]. It would appear that a market would exist in Australia for locally grown cork.

Plantations can be established from acorns or seedlings — both of which can be difficult to obtain. In order to produce quantities of good cork, the form of a tree can be manipulated by pruning and competition (spacing in plantations). The Glenloch Plantation has trees planted on a 2.4 metre grid and a 5.0 × 7.5 metre rectangular grid with an extra tree on the diagonal intersections [Banks et al., 1990]. The wider rectangular spacing and design is thought to be better for production and aesthetics [Banks et al., 1990]. The better the site conditions the better the growth and production [J. Banks, pers. comm.], with irrigation a possibility. Trees must be protected from fire and from grazing while young.

The first stripping of bark is done after 20—25 years and thereafter every 8—12 years. Trees continue to produce for several hundred years. The best cork is obtained from younger trees and is 2—3 cm in thickness [Cooke, 1961].

Conclusion

Although this summary paper has revealed the diversity of alternative tree and shrub product industries in Australia and the plant resources on which they are based, it has also shown that many potential industries have not flourished either because of their poor commercial viability or the lack of financial and research backing. There appear to be great disparities in the distribution and type of research afforded to these industries, with some having little or no research backing at all (e.g. craftwood) while others benefit from numerous research programmes (e.g. the eucalypt oil industry). Much of the research that is being carried out is focussed on technical aspects of the industry or the resource, while little has been done on economics or marketing. Currently most alternative tree and shrub product industries are not commercially viable as single enterprises primarily due to high production costs and overseas competition.

There is a demand for alternative tree and shrub products both overseas and in Australia. For alternative tree and shrub product industries to become viable, enterprises need to be integrated with other commercial activities, e.g. forestry, agriculture or other alternative tree and shrub products. Likewise, opportunities for enhancing farm income solely from alternative tree and shrub products are minimal. However, additional income could be made by incorporating appropriate species into multipurpose plantings.

This paper attempts to summarize a wide variety of alternative tree and shrub product industries in Australia and the possibilities for using them to enhance farm income. Although the authors' have attempted to incorporate as much information as possible, this report represents only a brief summary, and should not be viewed as comprehensive. We welcome any further information readers may like to contribute.

In conclusion we have drawn up a summary table of those alternative tree and shrub product industries which are currently, potentially or possibly viable for on-farm production, and the agro-climatic region where each industry is based (see Table 1).

Table 1. Summary table of alternative tree and shrub product industries.

Industry	Species	Region/climatic zone	On-farm viability
Wildflowers and foliage	*Banksia spp., Boronia spp., Dryandra spp. Stirlingia* spp., Kangaroo Paw, Waxes Verticordias	Dry temperate-semi-arid regions of Western Australia	Current
	Ixodia achilleoides, Helichrysum spp., Helipterum spp., Anigozanthos spp.	Dry temperate, semi-arid regions of South Australia	Current
	Aotus spp., Banksia spp., Callitris rhomboidea, Caustis spp., Eucalyptus spp., Gleichenia flaballatus, Leptospermum spp., Macrozamia spp., Melaleuca spp., Persoonia spp., Xanthorrhoea spp., Chamelaucium unciinatum, Helichrysum diosmifolium	Wet sub-tropics/tropics	Current or potential
	Blandfordia spp., Boronia spp., Telopea speciosissima, Doryanthes excelsa, waxes, kangaroo paw, *banksia* spp., leptospermum spp., foliage flannel flower (*Actinotus helianthi*), wax flowers (*Eriostemon* spp.) and *Crowea* spp.)	Wet-dry temperate central coast N.S.W.	Current-potential
	Eucalyptus spp., Helichrysum spp.	Semi-arid (north), dry sub-tropics/tropics	Potential
	Nothofagus cunninghamii, N. gunnii, Eucalypt spp.	Wet temperate rainforests of Tasmania	Current
	Telopea truncata, Thryptomene micrantha	Wet temperate regions of Tasmania	Potential
	Eucalyptus spp. leaves, *Xanthorrea* spp., Koala fern (*Restio tetraphyall*), *Boronia megastigma, Telopea speciosissima*	Wet temperate Victoria	Current
Nuts	Macadamia, *Macadamia integrifolia* and *M. tetraphylla*	Sub-tropical rainforest; Northern N.S.W. and Southern Queensland	Current
	Walnut *Juglans nigra/regia*	Wet temperate regions of Tasmania and Victoria	Potential

Table 1. (Continued)

Industry	Species	Region/climatic zone	On-farm viability
	Chestnut *Castanea sativa*	Wet temperate regions of Tasmania and Victoria	Potential
	Pistacio *Pistacia chinensis vera*	Semi-arid regions of Victoria and South Australia	Potential
Dates	*Phoenix dactylifera*	Queensland, NT	Current
		Victoria, saline semi-arid regions	Possible
Seed collecting	All species, mainly eucalypts	All states	Possible
Charcoal	River Red Gum (*E. camaldulensis*)	Along Murray River; semi-arid N.S.W./Vic	Current
	Eucalyptus anceps, E. dumosa, E. foecunda, E. gracilis, E. socialis, Casuarina stricta, C. leuhmanii, Acacia aneura	Arid and semi-arid zones of South Australia	Possible
Essential Oils	Brown Boronia (*Boronia megastigma*)	Western Australia	Current
	Eucalyptus dives, E. calycogona, E. augustissima, E. horistes, E. loxophleba, E. spathulata, E. kochii	Dry temperate, semi-arid Western Australia	Possible
	Melaleuca alternifolia	Northern rivers region of N.S.W. — wet temperate	Current
	Stringy bark peppermint *E. olida*	New England tablelands N.S.W.	Current
	E. polybractea	West Wyalong region, semi-arid N.S.W.	Current
	E. kochii	Condobolin region, semi-arid N.S.W.	Possible
	Sandalwood (*Santalum specalum, S. lanceolatum*)	Semi-arid, arid regions of South Australia and Western Australia and Queensland	Current
	Melaleuca alternifolia, M. linariifolia	Wet sub-tropics/tropics	Potential

Table 1. (Continued)

Industry	Species	Region/climatic zone	On-farm viability
Medicinal plants	Duboisia-hybrid between *D. leichhardtii* and *D. myoporoides*	Semi-arid regions of central and west Queensland, dry regions of Southern Qld	Current
	Solanum laciniatum	Tropical rainforests Qld	Potential
	Solanum laciniatum, S. aviculare	Wet temperate, subtropical regions	Potential
	Castanospermum australis	Tropical rainforests Qld	Possible
Christmas trees	*Pinus radiata,* Douglas fir, Dutch pine, Norway spruce	Wet temperate regions close to Melbourne, Victoria	Current
Honey	A variety of Australian most important individual sp. incl. *Eucalyptus* spp., Leatherwood (Tas.), *Eucryphia lucida, Banksia* spp. (WA)	All states	Current
Tannin	*Acacia pycantha, A mearnsii*	Semi-arid South Australia Victoria, N.S.W., Western Australia	Possible
	Pinus radiata	As a by product of pine plantations in wet temperate South Australia	Potential
	E. astringens	Semi-arid regions of south-west Western Australia	Possible
Quandongs	*Santalum acuminatum,* natural stands	Semi-arid regions of South Australia near Quorn	Current
	Plantations	Arid and semi-arid regions of Western Australia near Geraldton	Possible
Craftwood	Huon pine (*Lagarostrobas franklinnii*), King Billy pine (*Athrotaxis selanginoides*), Blackwood (*Acacia melanoxylon*), Myrtle (*Nothofagus cunninghamii*), Sassafrass (*Atherosperma moschalum*)	Wet temperate rainforests of Tasmania	Possible

Table 1. (Continued)

Industry	Species	Region/climatic zone	On-farm viability
	Jarrah (*E. marginata*), *Casuarina* spp., Blackbutt (*E. patens*), *Banksia* spp., Marri (*E. calophylla*), Yate (*E. cornata*)	Wet temperate forest and coastal heathlands of Western Australia	Possible
	River Red Gum (*E. camaldulensis*)	River red gum forests along Murray River; semi-arid N.S.W./Victoria	Possible
	Variety including: mulga (*Acacia aneura*), Acacia spp.	Arid semi-arid Queensland	Current/ possible
	Wide variety including: Red Cedar (*Toona australis*), Silky Oak (*Grevillea robusta*), Hoop Pine (*Araucaria cunninghamii*). Other rainforest species: *Eucalypt spp., Banksia* spp., Camphor laurel (*Cinnamomum camphora*)	Wet subtropical, tropical regions	Current/ possible
Broombush	*Melaleuca uncinata* from natural stands and plantations	Semi-arid regions of South Australia and Mallee-Wimmera region of Victoria	Current
Carob for Carob Powder	*Ceratonia siliqua* (irrigated)	Dry temperate to semi-arid regions Australia-wide	Potential

Acknowledgements

This paper would not have been possible to produce without the co-operation of many individuals who have assisted with providing specialist information. We are particularly grateful for the assistance of Penny Butcher, Russell James and Mike Slee from the ANU Forestry Dept. and John Doran from CSIRO Forestry who all assisted with the section on Essential Oils; to Suzette Searle of CSIRO Forestry who assisted with the section on Tannins; to Cynthia Carson and Jim Murison from Horticultural Research Stations at Redlands Qld and Gosford N.S.W. who assisted with the section on Wildflowers and Foliage; and to Andrew Beal of the Murray-Darling Basin Commission who assisted with the section on Quandongs.

Martine Scheltema from Greening Australia WA Inc., Ross Florence from the ANU Department of Forestry and Roslyn Prinsley from the Bureau of Rural Resources commented on previous drafts of the paper. Finally, I would like to thank Fiona Grant and Zosha Smith who helped with the typing.

References

Abbott PS (1986) Commercial production of oil from Eucalypt foliage. Proc For Prod Res Conf Melbourne, Vic

Allen A (1986) Growing Nuts in Australia. Night Owl Publishers, Shepparton, Victoria

Anderson H and Bilkens A (1958) Investigations on the Tannins of *Pinus radiata* bark. Aust J App Sci 9(2): 183—191

Applegate GB, Davis AG and Annable P (1990) Managing sandalwood for conservation in North Queensland, Australia. In: Proc Symposium on Sandalwood in the Pacific, April 9—11, 1990, Honolulu, Hawaii, pp 12—18. USDA Forest Service Gen Tech Report. PSW—122

Applegate GB and McKinnell FH (1991) The management and conservation status of Santalum species occuring in Australia. Paper to 17th Pacific Science Congress, 27th May—2nd June, Honolulu, Hawaii

Australian Bureau of Statistics (1990) Value of Agricultural Commodities Produced, 1988—89, Queensland. ABS Catalogue No. 7501.3

Australian Bureau of Statistics (1991) June Quarter. Foreign Trade. Australian Merchandise Exports/Imports 5434.0/5435.0

Australian Honey Board (1990) Annual Report 1989—90

Australian Tree Seed Centre (undated). Australian Suppliers of Tree Seed. CSIRO Division of Forestry and Forest Products

Banks JCG, Florence R, Taylor K and Egloff B (1990) Glenloch Cork Plantation. Report to ACT Office of City Management. ANUTECH, Canberra

Barton AFM and O'Reilly CM (1986) Eucalyptus oil project newsletter. School of Mathematical and Physical Sciences, Murdoch Univ, Oct 1986

Batt P (1990) Australian Native Flora Project. Preliminary report. National Resource Centre in Floriculture Marketing

Boland DJ, Brophy JJ and House APN (1991) Eucalyptus Leaf Oils. Inkata press, Melbourne

Brooker MIH, Barton AFM, Rockel BA and Tjandra J (1988) The cineole content and taxonomy of *Eucalyptus kochii* Maiden & Blakely, and *E. plenissima* (Gardner) Brooker, with an appendix establishing these two taxa as subspecies. Aust J Botany 36: 119—129

Brophy JJ, Boland DJ and Lassak EV (1989) Leaf essential oils of Melaleuca and Leptospermum species from tropical Australia. In: Boland DJ, ed, Trees for the Tropics: Growing Australian Multipurpose Trees and Shrubs in Developing Countries. ACIAR Monograph No. 10

Bulman PA, Scarvelis J, Mason BK and Wallace M (1991) Trees for other Products — South Australia. Paper to 'The Role of Trees in Sustainable Agriculture' Conference. Albury, 30th Sept—3rd Oct

Bureau of Flora and Fauna (1982) Flora of Australia, Vol. 29, Solanaceae. AGPS, Canberra

Castley M (1991) The role of trees in sustainable agriculture as minor forest produce: Tasmania. Paper to 'The Role of Trees in Sustainable Agriculture' Conference. Albury, Sept. 30th—Oct. 3rd 1991

Clemson A (1985) Honey and Pollen Flora. Dept Agriculture, NSW. Inkata Press, Melbourne

Colton RT and Murtagh GJ (1990) Tea-tree oil-plantation production. Agfact P6.4.6. NSW Agriculture and Fisheries

Cooke GB (1961) Cork and the Cork Tree. Pergamon Press, New York

Cribb AB and Cribb JW (1974). Wild Food in Australia. Fontana, Sydney

Cribb AB and Cribb JW (1981) Useful Wild Plants in Australia. Collins, Sydney

Doran JC and Matheson AC Genetic parameters and expected gains from selection for monoterpene yields in Petford *Eucalyptus camaldulensis*. New Forests (in press)

Elliot WR and Jones DL (1980) Encyclopedia of Australian Plants Suitable for Cultivation. Lothian, Melbourne

Esbenshade H and Wilson G (1986) Growing carobs in Australia. Goddard and Dobson, Box Hill, Victoria

Forestry Commission Tasmania, Division of Silvicultural Research and Development, Annual Report 1989/90

Harris S (1990) Three Tasmanian plants with potential for commercial development as sources of bush food. Paper to 5th Symposium of Australian Gastronomy, Adelaide

Hawkins BW (1991) Alternative tree and shrub products in NSW Paper to 'The Role of Trees in Sustainable Agriculture' conference. Albury, 30th Sept—3rd Oct

Heinjus D, A guide to farm tree planting in southern Australia. (in press)

Hillis WE (1987) Heartwood and Tree Exudates. Springer-Verlag Series in Wood Science, Berlin

Isaacs J (1987) Bush Food: Aboriginal Food and Herbal Medicine. Weldons, Sydney

Keally I (1989) Sandalwood. Landscope 4: 35—51

Lake J (1990) Medicinal tree crops have export potential. Aust Horticulture (July): 30—34

Lansdown C and Reid R (1991) Alternative tree and shrub products — Victoria. Paper to 'The Role of Trees in Sustainable Agriculture' Conference. Albury 30th Sept—3rd October

Lassak EV (1988) The Australian Eucalyptus oils industry, past and present. Chemistry in Australia (Nov): 396—398

Lassak EV and McCarthy T (1983) Australian Medicinal Plants. Methuen, Australia

Loneragon OW (1990) Historical Review of Sandalwood (*Santalum spicatum* Research in Western Australia. Research Bulletin No. 4, CALM, Como, WA

Maiden JH (1975) The Useful Native Plants of Australia. Compendium, Melbourne (Facsimile reproduction of 1889 edition)

Midgely S (1988) The Australian Tree Seed Centre: A Window to the Resource. CSIRO Division of Forestry and Forest Products

Minnis D (1990) Marketing horticultural products from Australia. Aust Protea Grower 2(3): 19—24

Murtagh GJ and Etherington RJ (1990) Variation in oil concentration and economic return from tea-tree (*Melaleuca alternifolia* Cheel) oil. Aust J Experimental Agriculture 30: 675—679

Newland N 'Broombush'. Master of Env Studies thesis, Adelaide University (in press)

Pearce L (1991) Tea tree miracle. Sunday Mail Magazine, March 17, 1991

Pegrum J (1989) Outlook for Floriculture. WA Dept Agriculture Misc. Publication 7/89

Pegrum J and Webb M (1990) An overview of the West Australian native plant and protea industries. J Int Protea Assn 19: 30—33

Robinson JB, ed, (1988) North Coast Agricultural Institute, Wollongbar Biennial Research Report 1986—1988

Ryan PA, Henderson RT and Applegate GB (1991) Potentially economic alternative tree and shrub products in Queensland. Paper to 'The Role of trees in Sustainable Agriculture' Conference. Albury, 30th September—3rd October

Scheltema MA (1991) Role of Trees in Sustainable Agriculture: alternative tree and shrub products in WA. Paper to Conference on 'The Role of Trees in Sustainable Agriculture'. Albury 30th Sept—3rd Oct, 1991

Searle S (1989). The forest phoenix: the rise and demise of the black wattle industry in Australia. Technical Paper No. 1, CSIRO Division of Forestry, Yarralumla ACT

Tasmanian Beekeepers Association Inc. (1990) Submission to Tasmanian Forest and Forest Industries Strategy Inquiry

Tasmanian Forest and Forest Industries Strategy (1990)

WA Department of Agriculture (1989) Proc Seminar on Eucalyptus Oil Production. Gnowanergup. WA Dept Ag Misc Bull 9/89

WA Department of Agriculture, Division of Agricultural Economics and Marketing (1989/90) Costs of production survey of Western Australian honey producers

Warren RG (1989) Drug plants — Australia's neglected industry. Aust J Biotechnology 3(1): 23—25

Windle BE, Barth GE and Lewis IR (1990) World Flower Market Overview. SA Dept Agriculture Technical Report No. 170

Wrigley J and Fagg M (1979) Australian Native Plants: A Manual for Their Propagation, Cultivation and Use in Landscaping. Collins, Sydney

Agroforestry Systems **20**: 59–86, 1992.

The role of shelter in Australia for protecting soils, plants and livestock

P. R. BIRD[1], D. BICKNELL[2], P. A. BULMAN[3], S. J. A. BURKE[4], J. F. LEYS[5], J. N. PARKER[6], F. J. VAN DER SOMMEN[7] and P. VOLLER[8]

[1] *Department of Food and Agriculture, Pastoral and Veterinary Institute, Hamilton, Victoria;* [2] *Department of Agriculture, Esperance Agricultural Centre, Esperance, Western Australia;* [3] *Woods and Forests Department, Adelaide, South Australia;* [4] *Department of Conservation and Environment, Natural Resource Management Systems Division, Melbourne, Victoria;* [5] *Department of Conservation and Land Management, Soil Conservation Service, Buronga, New South Wales;* [6] *Forestry Commission, Private Forestry Division, Kings Meadows, Tasmania;* [7] *Conservation Commission, Palmerston, Northern Territory;* [8] *Queensland Forestry Service, Extension Services Section, Landcare and Forestry Branch, Brisbane, Queensland*

Key words: windbreak, shelter, wind erosion, agroforestry, trees, sustainability

Abstract. The purpose of this review is to examine the current knowledge of the role of trees in providing shelter for pastures, crops, and livestock, for controlling erosion of soils and improving productivity and sustainability of agricultural production in Australia — and the extent to which this knowledge has been applied.

Land degradation — tree loss and associated soil salinity, water and wind erosion, soil acidification, soil structural decline and nutrient degradation — is evidence that our primary production systems are not sustainable. We have sought increased production without proper consideration of the ecological context of that system. About half of Victoria's crop and pasture lands are affected or at risk, and in Western Australia about 25% of the cleared agricultural land is wind-eroded and 60% is potentially susceptible, salinity affects 0.43 m ha and half of the divertible surface water is affected by salinity. Similar problems occur in other States. At least 43 m ha or 13% of our rangelands are seriously degraded by wind erosion caused by overgrazing, often coinciding with drought or a run of drier years.

'Minimum tillage' and stubble management for erosion control in cropping has been a major extension and research activity in Australian agriculture. Severe weather, combined with imperfect adoption of appropriate grazing and crop management systems, shows the weakness of complete reliance on these methods of erosion control. An effective system must accommodate the impact of extreme events, which are the most damaging. However, the complementary use of windbreaks to reduce soil erosion is rare, and their establishment has not been promoted, despite the wide-spread adoption of this technology by other countries.

In the cropping and higher rainfall grazing areas, the systematic planting of 10% of the land in a net of shelterbelts/timberbelts/clusters could achieve a 50% windspeed reduction; this would substantially improve livestock and pasture production in the short and long-term. Wind erosion could be dramatically reduced and crop production probably increased by the use of windbreaks. Wheat and oat yield at Rutherglen (Victoria), and lupin yield at Esperance (Western Australia), were increased in the sheltered zone by 22% and 47%, and 30%, respectively.

In semi-arid and dry temperate areas, planting of 5% of the land to shelter could reduce windspeed by 30–50% and soil loss by up to 80%. This planting would also contribute substantially to achieving other objectives of sustainable agriculture. Agroforestry — particularly timberbelts applications — will be important in the long-term strategy for achieving revegetation. If some of the trees yield a marketable product then the adoption of the system will be more readily achieved.

In the arid (pastoral) areas there is an urgent need to promote the ethic that preservation and improvement of the perennial grass and shrub vegetation is critical for the protection of the soil and maintenance of land capability. Control of animal grazing remains the sole means of preventing erosion in much of this zone. While satellite imagery allows us to assess the condition of leasehold lands, we have failed to achieve stocking policies that will halt the degradation of our rangelands.

1. Introduction

The purpose of this review is to examine the current knowledge of the role of trees in providing shelter for pastures, crops, and livestock, for controlling erosion of soils and improving productivity and sustainability of agricultural production in Australia — and the extent to which this knowledge has been applied. An attempt has been made, for each of the major regions in Australia, to determine:

a) the adequacy of existing shade/shelter for soil, plant and animal production
b) improvements in productivity and/or sustainability that can be achieved using trees/shrubs

and to identify crucial areas for further research or demonstration.

2. Tree cover — land degradation — sustainable agriculture

Two hundred years ago our forests covered 10% and woodlands 23% of the continent. Now 50% of those forests and 35% of the woodlands have been cleared or severely modified. Many of the remaining woodland trees are in decline and losses outstrip gains through planting [26].

Tree loss has been associated with almost every aspect of land degradation in Australia [36]; tree clearing for agriculture, and soil disturbance associated with grazing and cultivation, has produced massive soil erosion. This is most visible on steep slopes associated with sub-tropical horticulture or cropping where a tonne of crop product can result in the loss of 380 tonnes of topsoil [67]. In the wheatlands of Queensland the soil loss may exceed 50 t ha^{-1} y^{-1} (i.e. >4 mm y^{-1}), which is 50 times the most optimistic estimate of the rate of soil formation [27].

Retention of trees — and the re-establishment of more trees, assisted by appropriate farm planning is critical to the long-term objective of sustainable soils [18, 19, 24].

Tree clearing restrictions on freehold or leasehold land are currently in force in South Australia, Victoria and in a few water catchment areas in

Western Australia. In the latter state, remnant vegetation on 13 shires in the 300—600 mm rainfall zone ranges from 8 to 16% of farmland [5]. In New South Wales there are restrictions on clearing of trees in environmentally sensitive areas (as mapped by the Soil Conservation Service) on leasehold land in the Western Division, along water courses, and on land with slope >18 degrees east of the divide. No restrictions have been introduced in Queensland or Tasmania or the Northern Territory. The area cleared of trees in the Northern Territory is fairly small and largely the result of development of settlements, roads and services, firewood harvesting and concentration of stock around watering points [62]. Tree clearance in Queensland has not reached the extent of the southern states, although the pressure to do so is rising and there are significant areas of salinity and erosion. Mooted partial clearance of up to 50 m ha of poplar-box woodlands in Queensland may have serious adverse consequences resulting from the loss of an estimated 5 billion trees [26, 54]. Pasture development in that area, without further tree clearing, may be a better solution to the problem of long-term viability of cattle stations [54].

Land degradation — dryland salinity, water and wind erosion, soil acidification, soil structural decline, soil nutrient degradation — is Australia's major environmental problem [36]. About half of Victoria's crop and pasture lands are affected or at risk. Five million ha of land are affected by declining soil structure, 4 m ha are affected by declining soil fertility, over 0.2 m ha are salt affected and 25,000 km of gullies are actively eroding. The costs in terms of lost production and off-site effects are substantial. While these costs can only be generalised they already exceed $100 m per year and could be as high as $200 m per year [60].

The situation in Western Australia is no better: about 25% (4 m ha) of the cleared agricultural land is wind-eroded and 60% is potentially susceptible; water erosion affects 0.75 m ha; salinity affects 0.43 m ha and over 50% of the divertible surface water is affected by salinity; soil structure decline occurs on 3.5 m ha, sub-soil compaction on 8.5 m ha, acidification on 0.5 m ha and water repellence on 5 m ha [20, 55, 65]. The annual cost could be as high as $600 m.

Conservative estimates suggest that 43 m ha or 13% of our pastoral land is seriously degraded, with soil losses exceeding 20 t ha^{-1} y^{-1}. Compared with erosion of cropping lands or salinised areas the degradation of our rangelands is massive. The basic cause is overgrazing, often coinciding with major droughts or a run of drier years, and a reduced incidence of fire resulting in proliferation of woody shrubs at the expense of native forbs and grasses [67]. In spite of the best intentions, practices adopted in the past have not been adequate to contend with combined pressures of grazing and climatic variation of our pastoral dryland regions and degradation is continuing virtually unabated [32].

Management decisions taken to maximise output, such as the introduction to our harsh tropical rangelands of better adapted *Bos indicus* cattle, use of mineral and nitrogen supplements, and improved transportation have been cited as reasons for the present-day increased impact of grazing on our rangelands. We can also see politically inspired large-scale clearing of semi-arid land in the last thirty years, and a reluctance either to enforce compliance with current pastoral leasehold provisions or to review them and assist graziers to cope or leave.

Land degradation should be seen as a consequence of an inadequate agricultural or pastoral production system — a testimony to the fact that our systems are not sustainable — that we make decisions to improve primary production without proper consideration of the ecological context of that system [67].

3. Soil erosion

Most of our soils are based on very old weathered surfaces, without fresh additions of minerals from volcanic or glacial activity. Consequently, they are generally low in nutrients and erosion further depletes such impoverished material. Most of the nutrients are held in the organic matter of plants or in the top few centimetres of soil which is vulnerable to degradation following cultivation, or erosion by wind and water. Cropping and livestock production further depletes the nutrient pool as a result of export of minerals from the system in grain, meat and wool; losses for phosphorus, nitrogen or potassium are invariably not offset by fertiliser inputs or from weathering of rock [43, 67].

Tree replacement will be essential if we are to re-develop the fertility of many of our degraded soils and maintain sustainable systems of agriculture [36]. Agroforestry systems have an important role in conserving soil fertility [71]. Trees intercept rain and litter also reduces raindrop pounding of soils; surface litter increases soil organic matter, increases the water-holding capacity of the soil, promotes water absorption and decreases the rate of runoff; roots also bind the soil [47, 36, 71, 34].

3.1. *Soil erosion by water*

In Australia, more than 2.6 m km^2 (almost two thirds of the land used for grazing or cropping) needs treatment for land degradation [26]. Water erosion accounts for about 70% of the total Australia-wide [22].

Control of water erosion requires solutions based on whole catchment planning, incorporating pastures, cropping technology, grazing control, trees and physical structures such as contour banks. Trees alone are often insufficient to protect soil from sheet and rill erosion, since ground cover plants are also required to protect the soil surface [47, 71, 34]. In some circumstances shrub growth may prevent the growth of ground cover species and promote runoff. However, tree establishment in association with mulching or pasture is a practical means of controlling tunnel erosion on land with dispersible sub-soil, landslip on steep slopes and gully erosion [47, 24].

3.2. *Soil erosion by wind*

Wind erosion of agricultural or pastoral lands is a severe problem worldwide. Our dryland agricultural ecosystems are vulnerable to both natural and human impact.

The following examples from Western Australia are used to indicate the extent of wind erosion and the likely consequences to productivity [47]. Sheep grazing an 'average' sandy paddock in summer loosened 41 t of soil per ha and a wind storm removed 5 t of dust of the 28 t of topsoil lost per ha. A badly overgrazed paddock lost 30 t of dust in the 200 t of topsoil blown from each hectare of land. A clover pasture harvested for seed lost 18 t of dust in the 110 t of soil lost from the top 8 mm of each hectare. Losing the dust from the top 8 mm of soil reduced the yield of the next crop by 12—25%. It was also estimated that summer grazing on the same paddock for 7 years would reduce productivity by 50%.

Each soil has a particular propensity to erode — and sandy soils are particularly prone to wind erosion. The 'starting velocity' where sand flux begins is a function of soil type, pasture or trash cover, surface crusting from rain, soil particle aggregation (reduced by hoof pounding or cultivation) and moisture. Sand flux on a sandy soil may only begin when the windspeed at the surface exceeds around 9 m sec^{-1} but the rate of flux increases steeply thereafter as windspeed increases [6]. However, the fine fraction (clay, silt) has been removed from the surface long before the sand moves — the effect of the sand movement is to destabilise the new soil surface to allow further loss of 'fines' [53]. Loss of dust in smaller amounts is usually not regarded by the public as being soil erosion — yet the particles of clay and silt (< 50 microns) contain the bulk of the soil nutrients and organic matter and these are the first to move on windy days. Where cultivation or soil disturbance associated with grazing (hoof pounding/plant uprooting) reduces the aggregation of particles then loam or clay soils are also at risk.

We believe that the problem in Australia is as critical as in other continents and if we want sustainable farming systems then we should accept that the problem can not be overcome without reforestation. Current approaches in North America, the USSR and China are given below to indicate how other countries are dealing with soil erosion through revegetation — and to emphasise, by comparison, our neglect.

3.2.1. *Experience in North America, USSR and China.* In the Great Plains of America topsoil is being lost twice as fast as it was during the 'dust bowl' years [61]. Current USDA estimates are that 159 m acres of cropland are either environmentally sensitive or economically marginal for cropping; agroforestry options for this land would relieve the problem. The Center For

Semiarid Agroforestry is establishing a program of research, demonstration and technology transfer to integrate agroforestry and sustainable land-use systems. A major objective is to minimise loss of topsoil [61]. A windbreak model has been developed to assess alternatives for cropland protection in North America [39, 40].

In southern Ontario, Canada, sandy soils of glacial origin once devoted to dairy and beef cattle are now cultivated for soybeans, peas, tobacco, sweet corn and other field and processing crops; many single and double-row windbreaks of Cedar and Spruce have now been planted, despite extremely slow growth in this environment, to prevent wind erosion of the exposed soil. Others growing cash-crops on erodible sands are using a permanent-bed system with tall wheatgrass or cereal rye in beds about 12 m apart (8% of the area) to shelter the crop and soil.

Field shelterbelts in the USSR (2—4 row belts placed around 30 heights apart) reduce the windspeed in the protected fields by 30—40% and are regarded as an 'integral part of the agro-landscape' and 'the solution to the major task of rational utilisation and conservation of natural resources' [56].

In the European portion of the USSR there are 227 m ha of arable land (200—500 mm precipitation) and of this 90 m ha are subject to wind erosion. About 40 m ha of cropped land are protected by 2 m ha of shelterbelts and 3 m ha of woodlots; the combined effects of improved soil cultivation technology and shelterbelts have reduced average soil losses to around 20% of previous rates and in some regions very prone to erosion 0.8 m ha of two or three-row belts have eliminated wind erosion [56, 41]. The long-term target is for protective afforestation with shelterbelts of 2—2.5% of the arable lands in the forest-steppe, 3—4% in the steppe regions and 5—7% on light sandy soils and slopes — this will entail about 5 m ha of shelterbelts and 13 m ha of woodlots. Aerial and satellite photography are used to provide an inventory of shelterbelts, to study their effectiveness, to map and classify pastures and soils, and to plan protective works for erodible soils [41].

Trees were planted around cropping land in China to prevent wind erosion as early as 550 BC [59]. A return to that practice in the People's Republic of China began in 1950, with the planting of 4,000 km of forest belts to control sand movement on the 13% of China's land considered desert. Planting began in the 1960s of narrow forest belts along roads, canals and ditches, and small network grids on the farmlands. In 1978 the 'Three North' shelter-forest system was initiated to protect 12 provinces in semi-arid northern China extending 7,000 km east to west. The shelter-forest belts are mostly of 2—4 rows; the area within the grid varies with the climate but is 13—16 ha in the warm-temperate area of the sub-tropical zone. Within the grid there may be rows of deep-rooted small-crowned timber trees (e.g. Paulonia) planted 30—50 m apart to further protect crops.

After 40 years 30 m ha of shelter-forest has been established in China and the forest coverage on the agricultural plains regions (a fifth of China's cropland) has increased from 1.1% in 1950 to 7.8% [59].

3.2.2. *Wind erosion in Australia.* Wind erosion has been substantial across the Australian continent and particularly in the semi-arid lands in south-western New South Wales, South Australia, Victoria and Western Australia [53, 32]. However, the use of windbreaks to reduce soil erosion is rare and their establishment has not been actively promoted by any state or federal agency.

New South Wales. In the lease-hold Crown lands of south-western New South Wales the soils range from sand to clay and support vegetation ranging from mallee eucalypts to saltbush and spear grass. There is an embargo on clearing in the Murray geological basin and soils more sandy than sandy-loam may not be cropped. Loamy-textured soils may be at risk if cultivated or grazed bare. Wind erosion is not a major constraint to landuse for soils with a high clay content [53].

Victoria. Wind erosion in the Victorian Mallee has been very severe; large dust storms persisted into the 1950s [53]. After that period 'conservation farming' methods developed by the Department of Agriculture and promoted by the former Soil Conservation Authority and other bodies have reduced but not eliminated the problem [45]. Some 0.7 m ha of land are affected in the Mallee and Wimmera [60]. Predicted greenhouse effects on the Mallee and Wimmera areas of Victoria are increased windspeed and lower winter rainfall — both will increase erosion [53].

Despite conservation farming methods (stubble retention, minimum tillage, stock control) soil erosion by wind is still expanding [44] and in Victoria is still significant in drought years [60]. Spectacular events, like the day in February 1983 when an estimated 1—3 m tonnes of dust passed over Melbourne, are obvious effects of drought, combined with cultivation or over-stocking of erodible soils. The dust came from areas of southern Victoria and South Australia not usually associated with wind erosion. Ordovician sediments in the Wickliffe-Mortlake-Hexham area experienced topsoil and sand movement reminiscent of the Mallee in the 1930s.

The 1982 drought in southern Victoria was due to a shortage of rain between July and November inclusive (159 mm at Hamilton compared with the long-term mean of 369 mm), with a subsequent shortage of pasture growth that spring. Some of the damage in the summer and autumn of 1983 could have been avoided had farmers removed stock from the loam and sandy soils and confined them to clay soils. Where windbreaks were present the damage was reduced, but much sand was deposited to the windward and through the belt. There were insufficient belts to prevent the windspeed from recovering in the intervening space — and early failure to remove sheep from the paddocks left the soil in an erodible state.

Tasmania. Historically in coastal Tasmania and the islands, there has been actual or potential encroachment of sand-dunes onto agricultural land. A more recent development, however, has taken active erosion by wind into the north midlands, where clouds of soil are increasingly a feature of the spring scene.

The erodible soils are naturally occurring wind-blown sands in the extensive valley of the South Esk river. The economic imperative of the last few years, coupled with irrigation possibilities, has made this area more attractive for cropping of potatoes, cereals, peas and poppies. These soils have a like appeal for the processors, as an offset to the traditional vegetable growing krasnozem soils of the north-west, and through their prospects of clean root-crops and winter harvesting.

With recent annual precipitations much below the mean of about 625 mm and typically frequent high winds in spring and early summer, a multi-million dollar industry and a large soil resource are in jeopardy through over-exploitation and through inadequate protection by means of conservation cropping practices and appropriate shelter systems.

Western Australia. Most of the soils in the agricultural regions are susceptible to wind erosion; it was estimated in 1983 that about 25% of the 16 m ha of cleared agricultural land needed special attention and another 40% some management to control wind and water erosion [20]. Another report indicated that, in 1986, 0.5 m ha was actually affected by wind erosion, and another 10 m ha were susceptible [55]. 'Sandplain' soils (approximately 11.7 m ha, or 27% of the South West Agricultural Region) are particularly prone to wind erosion.

Anecdotal evidence of dust erosion in the central wheatbelt is widespread. Instances in 1991 include severe wind erosion in the north-eastern, central and south-eastern wheatbelt. The latter area, north of Esperance, was put at risk by the driest spring and summer on record. The central wheatbelt was affected by a decaying cyclonic low pressure system in May 1991, when many paddocks had been cultivated for crop establishment, and pastures were mostly bare.

Jerramungup, cleared mostly since the 1950s, has had significant wind erosion since 1969 and particularly in 1980, 1981, 1983 and 1984. About 0.44 m ha in 1980 and more than 0.64 m ha in 1981 were estimated to have been seriously affected by wind erosion [65]. This was about 7.3% of cleared land, and 18.3% of the cropped area showed evidence of sand-blasting.

Conditions leading to wind erosion are very similar in the Esperance agricultural region, about 300 km further east from Jerramungup, and based on the coast. Farming conditions, soils and winds experienced at times of high risk on the Esperance sandplain, have led to widespread and severe erosion.

'Minimum tillage' and stubble management in cropping have been a major extension and research activity of the Western Australian Department of Agriculture for nearly a decade. However, severe weather and/or economic factors, combined with imperfect adoption of appropriate grazing and crop management systems, show the weakness of relying on management systems developed for the 'normal' year. An effective system must accommodate the impact of extreme events — shelterbelts are a complementary erosion control aid.

4. Shrubs and trees for rangelands and low rainfall farmland

The greatest wind erosion occurs in the arid and semi-arid lands, where the pastoral industry exists entirely by a precarious exploitation of the natural vegetation [32]. The maintenance of this vegetation by animal management is the sole means of preventing erosion, but during drought when erosion is potentially the greatest this vegetation is regarded as a drought fodder reserve. The loss of perennial shrubs (including the saltbush species) mitigates against control of erosion, and soil losses are certain when arid lands are overgrazed. The drier the climate the greater the difficulty in matching livestock numbers to the vagaries of seasonal production. A single episode of mismanagement may cause serious deterioration from which recovery may be almost impossible [32]. Most of the arid and semi-arid lands are severely eroded as a consequence of the loss of vegetation. Only 10% of shrublands and 30% of the grasslands have been retained in a condition approximating the original state; 65% of the shrublands and 15% of the grasslands have degenerated to less than 60% of their pristine condition [32].

The pastoralist has, in the past, been encouraged by drought relief and subsidy arrangements to pursue a stocking policy based on the best years. Drought feed or transport subsidies have hitherto encouraged producers in semi-arid and higher rainfall areas alike to maintain these stock numbers during drought when the land is most susceptible to erosion. In contrast, prudent land managers did not benefit as much from their conservatism. The Federal government has recently reviewed its policy on drought, now recognising that it is a normal event and should be largely catered for by the producer.

In time we will show more interest in the growing of trees for timber on farms in the lower rainfall areas (< 600 mm). We suffer from the belief that forestry is only an option for high-rainfall regions and for short-rotation species. This myopia effectively limits our vision to 5—10% of the country and does nothing to promote the revegetation of the degraded remainder. For this attitude to change we should look again at forestry practises in Europe where one generation plants trees for the next to manage — and harvests those planted by previous generations. We need to adopt that tradition — but pioneers seldom reap the reward of their enterprise.

It is possible to incorporate eucalypts (firewood, posts, poles), acacia (posts, decking, furniture), callitris (decking, furniture), casuarina (shingles, furniture), melaleuca (brushwood), sandlewood (scented wood), quandong or carobs (fruit) and other species in cluster planting or timberbelts. Whilst the trees are relatively slow-growing under low-rainfall conditions, many yield high value products and their inclusion in shelter networks will be of long-term benefit, particularly where they are managed to permit regeneration. In Portugal the evergreen oak *Quercus suber* is grown for shade, to improve grass production (through shelter, increased soil organic matter from leaf fall, and leaf drip), to reduce soil erosion (largely from storm runoff) and to produce cork [37]. Cork is stripped after 30—40 years then perhaps every 10 years thereafter.

Despite slow growth rates, we could argue that in many cases forestry provides a better option for sustainable multiple-use of leasehold lands than the current pastoral activities which have caused such devastation.

5. Wide-spaced trees and clusters of trees for farmlands

Isolated groups or individuals lend a pleasing park-like appearance to the landscape (for example, the river red gum (*E. camaldulensis*) country of western Victoria, or the white gum (*E. wandoo*)/red gum (*E. marri*) country of SW Western Australia) and these provide much needed shelter. Mature river red gums (17 trees/ha) in a paddock at Vasey (western Victoria) showed that the trees reduced the wind speed at 1.5 m above ground to 50—60% of that in an adjacent open paddock [38].

Trees obviously reduce the paddock area that can be cropped. Estimates of this for both *E. wandoo* and *E. marri* near Arthur River in Western Australia, are 0.006% per tree per ha, with a further 1.8—2.4% reduction per tree per hectare due to competition with the crop within 20 m of the tree (T.R. Negus, pers. comm.). However, estimated yield losses such as the latter are difficult to substantiate because they take no account of the effect of trees in modifying the microclimate across the paddock (increasing crop yields in areas away from the trees) or effects on the site fertility (nutrient cycling and root-soil interactions). The long-term effect of clearing the trees on that site may be to depress crop yield.

Oliver Guthrie, a wheat farmer near Donald in the eastern Wimmera of Victoria, has retained about 70% of his original bulloaks (*Allocasuarina luehmannii*) and river red gums, with clumps and scattered trees averaging about 2—3 trees per ha. He has intensively cropped the area since 1950 and believes that the gain from trees in sheltering the crop and field from erosion outweighs the loss around each tree [70].

We do not have good data in Australia of the effects of tree species and spacing on pasture growth. There is anecdotal evidence that certain species are particularly competitive (e.g. *E. sideroxylon*) whilst other species, such as *E. camaldulensis* and *Casuarina cunninghamiana*, allow good growth up to the base. This effect is very noticeable around old river red gums on old or newly sown pasture in the 600—750 mm rainfall area of western Victoria — the pasture remains greener for a longer period and may be more vigorous.

Data for several species will be forthcoming from spacing-design projects at Hamilton in the next few years.

Cluster planting (small fenced woodlots dispersed over the landscape) is an economical means of providing shelter for livestock whilst preserving the open woodland character of the property and indigenous elements of the flora and fauna. Without such protection it will be virtually impossible to maintain the existing biotic elements; most of our grazed farmlands have only sparse mature trees, without any seedling regeneration, and the price of planting and protecting individuals from livestock is so high as to be prohibitive.

6. Shelterbelts for farmlands

Many of the farmlands have been extensively cleared and the most economical and practical option for revegetation, particularly in cropping country, is to provide trees and shrubs in belts which may include timber species for future harvest.

6.1. *Effect on windspeed*

Research in Victoria [11] has confirmed very extensive overseas research [3, 16, 29, 57, 58] that a variety of shelterbelts afford significant protection from wind. Some data are given in Table 1.

Some observations from this study are:

1. Despite significant differences in tree density all belts give a substantial degree of protection across the paddock, actual distance protected being proportional to tree height. Thus, taller belts give greatest distance protection — almost 500 m for a 20 m tall belt of *Pinus radiata* (No. 7) compared with about 250 m for a 10 m tall belt of *Cupressus macrocarpa* (No. 4). This is an important practical consideration — belts planted for soil protection should include species that are long-lived and will attain a substantial mature height. Retention of isolated mallee strips will provide some landscape protection, and preserve elements of the regions flora, but complementary planting of a taller species adjacent to the mallee would improve the shelter.
2. As indicated in Table 1, even a single row of tuart (*E. gomphocephala*) can make a large impact, despite the sparseness of foliage and the 4 m gap between foliage and ground. Fencing and including shrubs in the belt would eliminate wind gaps and improve the effectiveness of the belt, but in many instances this is not achievable. Too much attention has been paid to density aspects — we strive to achieve the impossible and yet the structure that we can achieve provides a significant wind protection to soil, plant and livestock. Our main objective must be to encourage the

Table 1. Effect of windbreaks on windspeed reduction in south western Victoria. Windrun was measured at 0.5 m and/or 1.5 m above ground (G) and the result expressed as % of values at the respective levels (G) in the open.

Species	Belt type and width	Height (m)	Trees/m of belt	G	Windspeed (% of open values) at various distances (H) from the belt: Windward −3H	Windward −H	Windward 0	Mid belt	Leeward 0	Leeward H	Leeward 3H	Leeward 6H	Leeward 12H	Leeward 24H
1. Sugar gum	Sown, 20 m, nfd	18	7	1.5	78	72	73	–	74	66	47	37	55	76
2. Sugar gum	Sown, 28 m, 4 rows, nfd	19	4	0.5	85	106	97	–	130	86	49	43	73	100
				1.5	85	76	76	–	107	76	49	39	72	91
3. Tuart gum	Single row, nfd	15	0.43	0.5	96	72	–	114	–	92	53	52	62	98
				1.5	95	81	–	108	–	110	73	61	77	100
4. Monterey cypress	Single row, fd, 4.5 m	10	0.29	1.5	89	80	–	56	–	58	37	30	54	82
5. Monterey cypress	Single row, nfd	15	0.17	0.5	85	58	–	133	–	102	49	29	(78)	(100)
				1.5	86	61	–	92	–	80	50	27	54	(84)
6. Monterey cypress	Single row, fd	9	0.25	0.5	89	78	–	81	–	68	37	33	71	100
				1.5	88	66	–	55	–	(49)	45	24	52	97
7. Monterey pine	Two rows, 6 m, nfd	20	0.33	0.5	85	66	123	–	116	90	65	48	54	82
				1.5	62	47	45	–	45	40	37	45	74	95
8. Young acacias	Four rows, 7 m, fd	3	2	0.5	100	–	–	–	–	20	42	68	89	–
				1.5	100	100	95	–	26	23	46	66	83	(91)

fd = fenced; nfd = not fenced; H = height of belt; ( ) = values by extrapolation.

planting of those species that will survive and grow in some very stressful environments.

3. Gaps beneath windbreak trees result in acceleration of windspeed at that point (No. 3, 5). With the unfenced cypress (No. 5) cattle had browsed the lower 2 m of foliage and the windspeed measured 50 cm above ground was 33% greater than in the open. Clearly this could create a significant unseen hazard (e.g. erosion, exposure of sheltering stock, accelerated passage of a fire front). A similar effect has been shown by others to occur at the ends of belts [29] and this should be countered by tapering off the ends with shrubs and grass. Belts should be fenced to retain lower foliage and shrubs.
4. Maximum wind protection for the wind-permeable belts occurred at around 6H to lee. There was still substantial protection at 12H (windspeed 50—80% of open values). Note that the impermeable belt (No. 8) which consisted largely of acacia species with sheoaks (*Allocasuarina verticillata*) and phalaris 2 years after planting, had maximum protection at 1H from the belt. However, at age 7 this belt has now become more permeable as the wattles and sheoaks developed and would conform more with the others. For some purposes, such as protection of freshly shorn sheep from rain and wind, a small area of dense shelterbelt is desirable [12].
5. It is apparent that belt profile is not of great importance. A myth has been perpetuated in the extension literature that a sloping profile is required and obtained by planting shorter species on outside rows. Evidence from the sugar gum (*E. cladocalyx*) and pine belts (Table 1) and shelterbelts in New Zealand [57, 58] confirm early work that a uniformly permeable vertical face is a desirable structure [16].

For practical purposes it may be necessary to site shelterbelts at around 10—15H apart (at maturity) to obtain the most economic response in terms of pasture and animal production [10]. Shelterbelts spaced at about 12.5H would also greatly reduce soil erosion, since windspeed in the area between the belts would be reduced by an average of about 50%. Siting belts at intervals of around 25H could also reduce windspeed by an average of about 33% in the sheltered gap.

6.2. *Control of soil erosion*

Wind erosion (E) is proportional to the wind speed (W) cubed for loose sandy soils, when wind velocity is above that required to initiate erosion [2, 42].

$$E \propto W^3$$

Small reductions in wind speed would, therefore, result in large soil erosion control benefits — this is an important point that is largely unrecognised by our agricultural advisers and land managers. For example, windspeed 0.5 m above the ground at 12H beyond the seemingly inadequate single-row tuart shelterbelt (Table 1) is reduced to 62% of open windspeed, but the estimated erosive force of the wind is reduced to 24% of open values. This situation is illustrated in Fig. 1. This model also shows the effect of the large gap in the foliage beneath the tree crowns — on susceptible soils wind erosion in this zone will occur unless ground cover is maintained by fencing to exclude livestock.

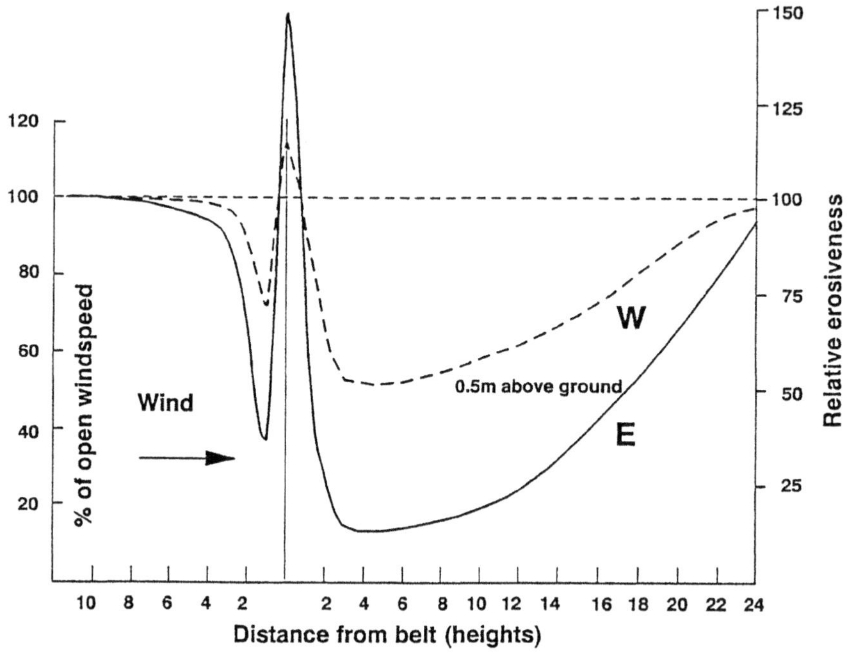

Fig. 1. Effect of a single row of Tuart trees (15 m tall, 2.3 m apart) on windspeed (W) at 0.5 m above ground and predicted relative erosiveness (E) of wind at varying distances from the belt.

Soil flux increases with windspeed and the area protected by the windbreak is less at higher windspeed [42]. The area protected also decreases as the wind approach angle becomes more acute to the windbreak [30, 31]. When the wind is parallel to the windbreak there is some erosion protection but maximum soil fluxes can be expected to occur beyond 8H (Fig. 2a).

When the wind is at 45 degrees to the windbreak (Fig. 2b), only the highest windspeed, 21.5 m s^{-1}, results in maximum soil flux at 24H. All other wind velocities have lower than maximum soil fluxes and erosion control (<5 g^{-1} m^{-1} s^{-1}) was maintained to 24H for all winds less than 14 m s^{-1}.

With the wind at 90 degrees to the windbreak, erosion control is main-

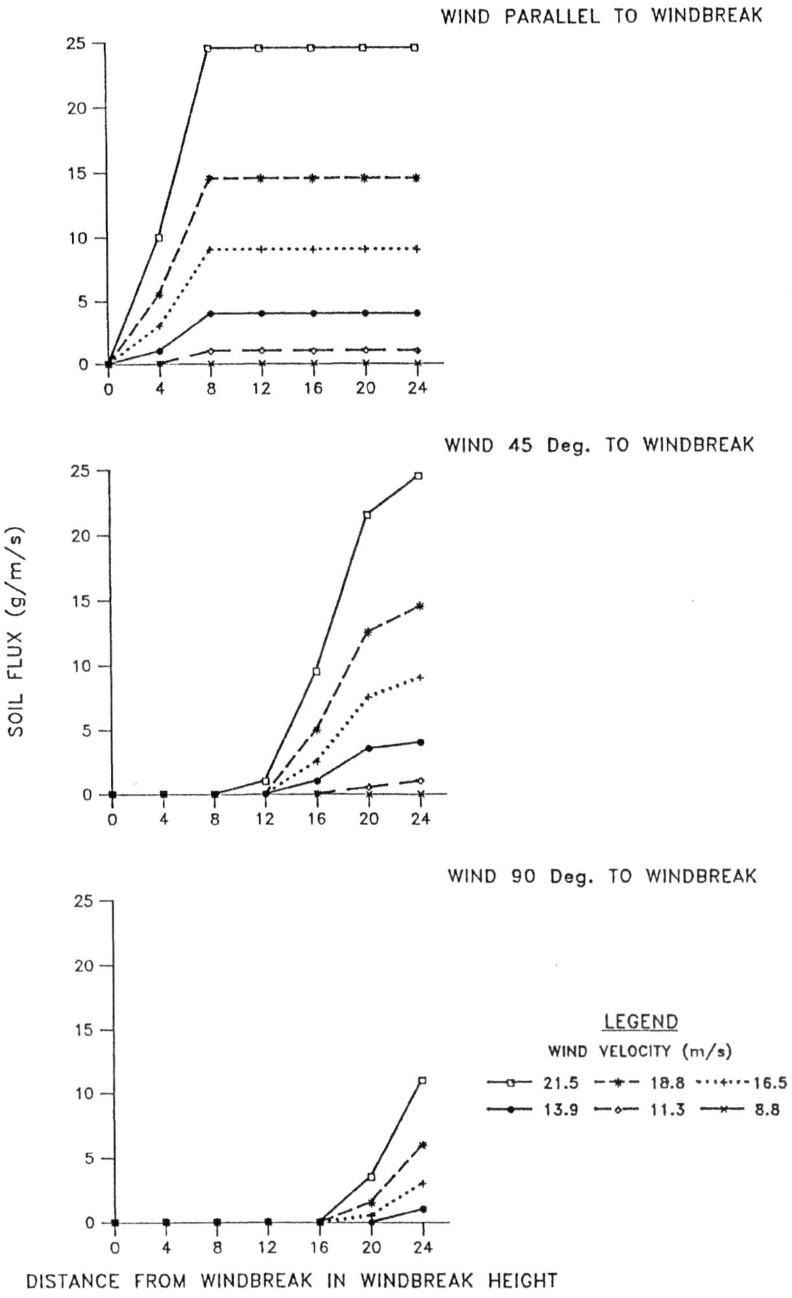

Fig. 2. Predicted soil flux to windward of a 40% porous windbreak.

tained to 24H for winds up to 18 m s^{-1} (Fig. 2c). Extrapolation of the data indicates that maximum soil flux rates would not occur until approximately 30H.

Tree windbreaks for the prevention of wind erosion would overcome deficiencies in current short-term agronomic management for protection during severe conditions. Combinations of tree and agronomic management systems are likely to be the most effective over the entire range of risk conditions.

Shelterbelts to reduce windspeed over vulnerable paddocks would substantially reduce wind erosion and loss of fertility, but what are the practicalities?

The main determinants of windbreak network development are the likely mature height of the windbreak, the distance apart of belts, the positioning of gaps, and the width of belt needed to achieve satisfactory shelter [29]. Only 2% of land would be occupied by belts that were 20 m tall at maturity, 10 m wide and spaced 25H (500 m) apart. If trees are planted in rows 3 m apart, with outside rows 2 m from the fence, 3 rows could be planted in a 10 m belt. In areas of southern Australia where annual rainfall exceeds 400 mm this could be achieved with belts of *E. cladocalyx* (sugar gum), *E. occidentalis* (flat-topped yate), *E. camaldulensis* (river red gum), *E. microcarpa* (grey box), pinus or other species suited to the site. Conversely, for mallee belts only 5 m tall, 7.4% of land would be used if the 10 m wide belts were spaced 25H (125 m) apart [11]. Belts wider than 10 m are required when 4 or more rows are planted; when the spacing between rows is 4—5 m (this may be necessary in semi-arid areas); when strips of remnant vegetation are to be retained in a viable condition; when more than 2 m is left between the fence and the outside rows of trees. For 20 m wide belts the percentage of land devoted to trees in the above cases is 4 or 15%.

We consider that appropriately positioned windbreaks offer major advantages for reducing wind erosion. This conclusion is supported by results and widespread adoption of windbreak technology in the USSR [56], USA [30, 61] and China [59]. We believe that further research in Australia will show that windbreaks for soil erosion control should not be considered 'impractical on most rural properties' [47]; on the contrary, they will be regarded as a necessary adjunct to sound pasture, stock and crop management practice on most rural properties [1].

7. Shelter effects on plants and animals

The many effects of trees on plant and animal physiology and production have been extensively reviewed in recent years [6, 7, 9, 10, 13, 51]. A summary of the important effects of shelter are presented below.

7.1. *Trees, shelter and plant production*

Trees may affect pasture or crop production in the following ways:

1. by reducing water loss, as a result of reduced windspeed and/or shading — this can also prolong pasture growth in summer [7, 51]. Pasture growth may also be increased by provision of shelter in the irrigation areas, with added benefits of a reduced loss of applied water and increased crop water use efficiency [7, 13, 51].
2. by retaining the heat of soil and air in the sheltered areas in autumn-winter-spring [29, 56, 72], and preventing overheating from dry hot winds during summer [59], crop yields are increased or the growing range extended.
3. by reducing inward radiation during the day and decreasing outward radiation at night [29] — it is a common observation that trees afford ground plants protection from frost. This has been observed experimentally in the sub-tropics where the grass remained green in winter under the trees [17]. Frosting may occur on the up-slope side of a belt on hillsides, unless some air drainage is provided [29].
4. by protecting soil by leaf or twig litter and/or reduced windspeed at the soil surface — reducing erosion of soil nutrients by wind and water runoff [71].
5. by competing with pastures or crops for light or soil moisture [6, 17, 50]. Photosynthetic rates of corn are maximised 10H from shelter but are low immediately adjacent to the trees [72]. Shading affects pasture species differentially. Thus, growth of Nangeela cultivar of sub clover and Berseem clover is reduced by only about 10% but other legumes by 40% under 50% shade [50]. Maku lotus also appears to cope with reduced light. Inevitably there is a point at which increasing density of trees completely excludes pasture. This point, and the shape of the response curve, will probably vary for different tree and pasture species. With some combinations it is possible that pasture production could be greater with a certain density of trees than without trees. We do not have good data on these interactions.
6. by reducing mechanical damage to plant tissue arising from flexion or sandblasting of seedlings [6, 7, 13, 51, 72].
7. by promoting mineralisation of soil nitrogen as a result of shading pasture or soil — and thereby increasing pasture growth [17, 68].
8. by contributing to soil organic matter (leaf and twig litter forming humus) and improved soil moisture retention [48, 56, 71, 72].
9. by improving soil physical properties, associated with soil root mass [71].
10. by trapping or recycling nutrients — these attributes may permit sustained yields from land where cropping rapidly depletes the soil nutrients in the absence of fertiliser applications [71]. Nitrogen fixation by acacias,

casuarina and other shrubs; trapping of dust on foliage and deposition of nutrients to ground with rain; accessing nutrients by root-induced rock weathering in the B and C horizons; recycling of minerals leached beyond the A horizon are all possible modes [4, 6, 48, 56, 71, 72].

11. by reducing or preventing soil acidification — acidification is a particular problem on soils having a low initial base status and where cropping and/or grazing with legumes is practised. Most pasture and crop species grow poorly on acid soil — some minerals are rendered unavailable (e.g. P), or too soluble, leading to Al or Mn toxicity or loss of N, Ca, Mg and K through leaching [33]. Such acidification is aggravated when the N produced from pasture or crop legumes (e.g. sub. clover) is converted to NO_3^- and leached from the root zone before it can be re-converted to plant or microbial protein. This is more likely to happen when there are few perennial plants. A challenge for the future is to develop production systems that efficiently recycle nutrients released from weathering or applied as fertiliser, and that minimise nitrate leaching [33]. The role of trees has still to be widely recognised, but there is evidence that trees in pastures have prevented soil acidity increasing [21], possibly through their trapping of nitrate and by their substantial additions of Ca in leaf drip and litter return [6].
12. by reducing deep infiltration of water to groundwater systems — and thereby preventing the rise of saline watertables which would otherwise result in deposition of sodium and other salts at the surface, with consequent depression of pasture and crop growth [35, 56, 59, 60, 64].

Windbreaks may assist horticultural production in following ways:

1. by preventing rubbing damage to fruit — e.g. citrus, stone-fruits and tropical fruits such as mangoes — the damage is largely cosmetic but may lead to the down-grading of 30% of crops in some areas of Queensland [63].
2. by preventing mechanical damage to orchards — e.g. protection of macadamia plantations in Queensland [63].
3. by controlling disease and insects — windbreaks may assist in the control of anthracnose outbreaks in mangoes [63] and as alternative hosts for leaf-feeding red shoulder beetles (*Monolepta*). These beetles are attracted by *E. torelliana* trees and can then be treated with insecticide away from the fruit trees. Windbreak trees also afford habitat for predatory birds and insects which may help control some insect pests.
4. by improving pollination efficiency — e.g. fruit set of cucurbit crops in Queensland may be increased 30% in some years [63].
5. by controlling spray drift — shelterbelts may be used to reduce the drift of chemical sprays from cotton fields; the windbreaks also extend the opportunities for spraying in otherwise unsuitable weather conditions [63].
6. by improved production from shelter — some varieties of apples and

cherries are tolerant of wind but other stone-fruits are very responsive to shelter [65].

7.2. *Trees and effects on animal production*

Trees may affect animals in the following ways:

1. by indirect effects on animals through effects on pasture production (see 7.1)
2. by providing foliage or fruit which supplement their intake from pasture. Examples of indigenous vegetation used for this purpose include mulga (*Acacia aneura*), saltbush (*Atriplex spp.*) and kurrajong (*Brachychiton spp.*) [32]. Useful introduced species include tagasaste (*Cytisus polmensis*) to provide green leaf in the summer, paulownia (*Paulownia spp.*) and poplar (*Populus spp.*) which provide edible leaves, sub-tropical shrubs such as leucaena (*Leucaena leucocephala*) which provide browse or cut foliage, and carob (*Ceratonia siliqua*) or honey locust (*Gleditsia triacanthos*) which produce edible pods [52].
3. by reducing livestock maintenance requirements due to shelter, as energy expenditure is increased by excessive heat or cold. This diverts energy from productive purposes into maintenance functions. It may also alter grazing behaviour and thereby reduce intake. Experiments with penned sheep and cattle indicate that strong wind and rain can double the requirement of energy for maintenance. A 33% reduction in windspeed from 3 m s^{-1} (10 km h^{-1}) to 2 m s^{-1} can result in a 10% saving in energy; a 55% reduction on windspeed would increase that value to 17.5% [6, 7, 51].

   Cattle appear to be particularly sensitive to hot humid conditions and the provision of shade on rangeland or dairy farms improves their tolerance and level of production [6, 23]. Shade also improves utilisation of pasture on rangeland. Provision of shade to cattle in feedlots in Queensland has improved feed conversion efficiency and survival [63]. Recent well publicised incidents where hundreds of cattle perished in unshaded feedlots in Queensland have resulted in prosecution of the operators. Cattle in unshaded saleyards in Victoria have also perished during very hot weather.

   It is unlikely that sheep with fleeces suffer significantly from heat stress in temperate regions.
4. by shelter effects on fertility. Heat stress can reduce ram fertility and, particularly with shorn sheep, affect ovulation, oestrus, conception and embryo survival. Heat stressed cows produce smaller calves and have longer intervals between calving. These effects with ewes and cattle are more pronounced in the tropics and sub-tropics but ram infertility as a consequence of heat stress is not uncommon in temperate regions [51]. The foliage of some tree species may cause abortion in cattle (e.g. *Cupres-*

sus spp.) while death may result from ingestion of foliage of sugar gum shortly after lopping.

5. by shelter effects on newborn animals. Published studies in SE Australia show that lamb mortality is decreased by an average of 50% when adequate shelter is provided [7]. Losses are worst in windy, wet and cold weather — perhaps all lambs born in such a bout will die. However, if wind speed can be reduced by 50% (e.g. from 3 m s^{-1} to 1.5 m s^{-1}) during such weather the predicted effect is a 50% increase in lamb survival [25]. On average, lamb losses can be reduced from around 20% to 10% of those born alive [7]. Since ewes will not always seek shelter prior to lambing, the shelter must either be dispersed (e.g. woodland with logs, tussocks etc.) or the sheep need to be confined to a sheltered area for maximum effect [12, 69].

   Provision of shade can improve the survival prospects of calves and lambs. On shadeless rangelands most of the 5—10% calves that die in the first week do so from heat stress [23]. Provision of shade in lambing paddocks in the sub tropics also improves lamb survival and subsequent growth and wool production [6, 7, 51].
6. by shelter effects on newly shorn sheep. Losses of shorn sheep following cold, wet and windy weather in southern Australia can be catastrophic [6, 9, 12, 51]. The conditions leading to the losses — and the use of shelterbelts and havens to reduce or eliminate them — have been discussed in the above papers. Regrettably, while most farmers are well aware of the danger, many have not made an effort to rectify the problem. A shelter haven can be created inexpensively by direct-seeding of trees of mixed species — and will be effective within two years of sowing, as evident from blocks established by the Department of Agriculture around Hamilton, Victoria. A brochure is available which illustrates several options for havens [69].

7.3. *Shelter effects on agricultural production*

Substantial effects of shelter on agricultural production have been claimed by those countries in which extensive shelter networks have been developed. Over a 25 year period from 1955—1980 in the USSR the average yield response of crops to shelter (allowing for land occupied by the belts) was: cereals 18—23%, industrial crops 20—26%, forage crops 29—41%, citrus 20—26% [56]. In China, protective effects are seen up to 20H from belts. Crop yields in 500 m × 500 m shelter networks in north-east China, during the years 1986—90, increased by 9—32% [72]. Another report of long-term effects of shelter in China shows increases for wheat, barley, rice and corn of 10—25%, 6—14%, 5—15% and 20%, respectively [59].

A case has been made for the deployment in the higher rainfall (>600 mm) grazing areas of southern Australia of 10% of the land in a shelter network [9, 10]. The model predicted that, for discount rates less than about

7% real interest, it was better to have 10% of the farm in a network of close-spaced belts (12.5H apart) rather than 5% of the farm in wide-spaced belts. The economics of the system would be further improved by the incorporation of timber species and management of the timberbelts [8]. While the simple models above give support to the economic feasibility of a farm revegetation program we need to use a sophisticated model, such as the Farmtree model [28], when considering a particular species, site and layout for a chosen option.

'Farmtree' [28] is an agroforestry simulation model for micro-computers in which details of site, layout, species, intended harvest age, proposed management and other data are entered and likely costs, effects on agriculture, tree growth, timber value and direct effects on farm income and net rate of return are estimated.

A preliminary model for a lower rainfall area in South Australia (450 mm) suggests that tree establishment to provide shelter or control salinity would also be profitable in the long term. With this model it is possible to make an allowance for land protection values as well as short-term agricultural production factors [14].

We require substantive evidence to verify the assumptions made in the models, particularly in relation to the effects of shelter on plant production (crops and pastures) for a wide range of Australian conditions. However, we now have evidence from Rutherglen, Victoria [15] that shelter provided by eucalypts can increase wheat and oat crop yields in the sheltered zone (1.5–

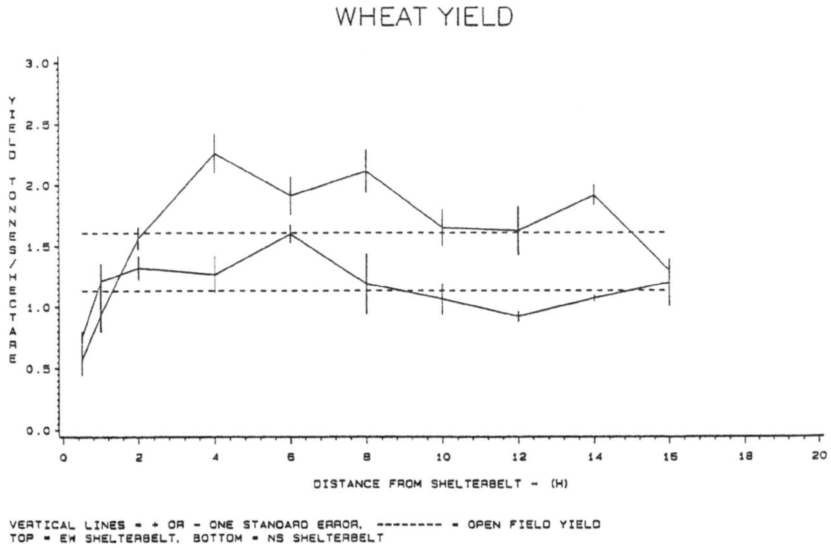

Fig. 3. Effect of shelter from east-west (EW) and north-south (NS) orientated belts on wheat grain yield at Rutherglen, Victoria.

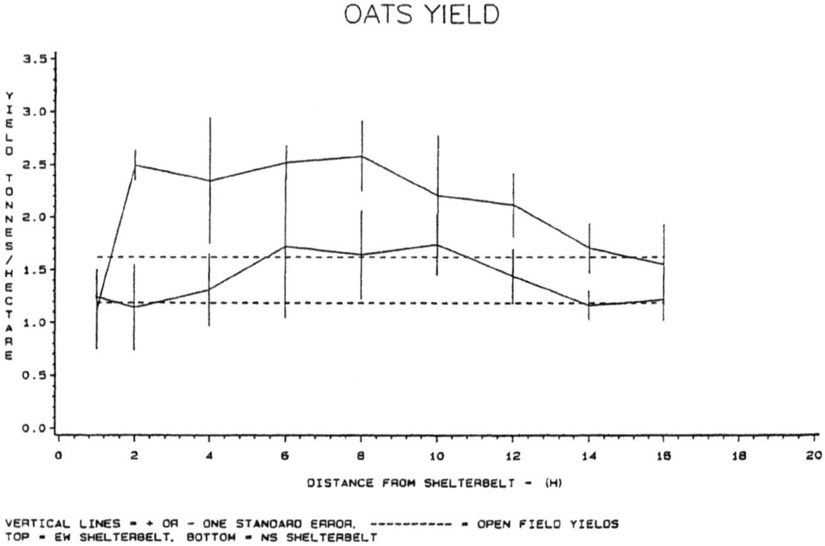

Fig. 4. Effect of shelter from east-west (EW) and north-south (NS) orientated belts on oat grain yield at Rutherglen, Victoria.

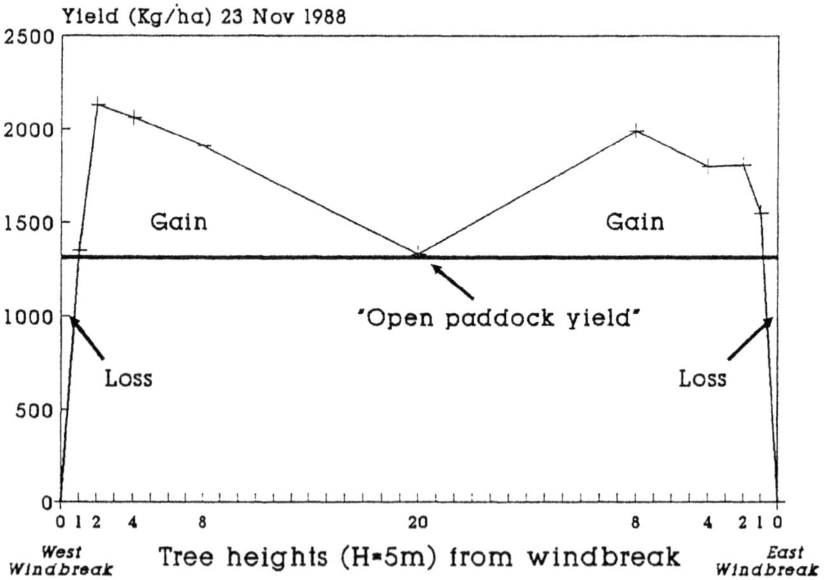

Fig. 5. Lupin grain yield between parallel pine windbreaks at Esperance, Western Australia, 1988 (property of G & J English).

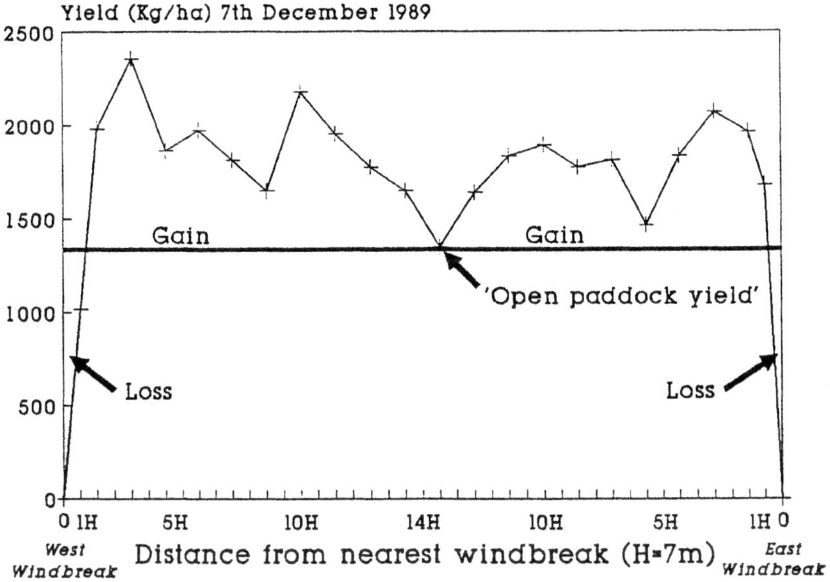

Fig. 6. Lupin grain yield between parallel pine windbreaks at Esperance, Western Australia, 1989 (property of G & J English).

9H) by 22% and 47%, respectively (Figs 3 and 4). Also, work at Esperance in W.A. [5] shows that lupin yields have increased by up to 30% when sheltered by *Pinus radiata* windbreaks (Figs 5 and 6). These data confirm experimental results world-wide [3, 6, 9, 13, 16, 29, 56] and we should procrastinate no further in our advocacy of the use of shelterbelts in Australia.

8. Conclusion

In the cropping and higher rainfall grazing areas, the systematic planting of 5—10% of the land in a net of shelterbelts/timberbelts/clusters could achieve a 50% windspeed reduction; this would substantially improve livestock and pasture production in the short and long-term. Wind erosion could be dramatically reduced and crop production probably increased by the use of windbreaks.

In semi-arid and dry temperate areas, planting of 2—5% of the land to shelter could reduce windspeed by 30—50% and soil loss by up to 80%. This planting would also contribute substantially to achieving other objectives of sustainable agriculture. It has been estimated, for example, that around 25% of the 300—700 mm rainfall belt in Western Australia will need to be reforested in order to prevent salinity from increasing ten-fold to 2.5 m hectares within 20 years [66].

In the arid (pastoral) areas there is an urgent need to promote the ethic that preservation and improvement of the perennial grass and shrub vegetation is critical for the protection of the soil and maintenance of land capability [32]. Control of animal grazing (introduced and native) remains the sole means of preventing erosion in much of this zone. While satellite imagery allows us to assess the condition of leasehold lands, we have failed to achieve stocking policies that will halt the degradation of our rangelands and revegetation strategies to restore them.

Agroforestry — particularly timberbelts (where timber and shelter are products) — will be important in the long-term strategy for achieving revegetation. Whilst trees are relatively slow growing under low-rainfall conditions, many yield high value products and their inclusion in shelter networks will be of immense long-term benefit. If some of the trees yield a marketable product then the adoption of the shelter system will be more readily achieved. Regrettably, forestry has only been considered an option for high-rainfall regions and for short-rotation species. We need to look at attitudes towards forestry in Europe, where one generation plants trees for the next to manage — and harvests those planted by previous generations.

We have the technical means of achieving goals of 10% or more of the land devoted to tree cover and the shelter and other benefits that follow. To that end, direct-seeding of trees and shrubs will increase the speed of revegetation and markedly reduce costs. This technology is well developed in Victoria, South Australia and Western Australia, and has gained much acceptance among farmers. What is required now is the political will to vest with some meaning the new bureaucratic jargon 'sustainable development'.

There are a number of areas in which research must be improved if we are to positively model biological and economic responses of soils, plants and animals to shelter:

- the effects of windbreaks in combating soil erosion should receive high priority in the arid, semi-arid and dry temperate zones — unlike other countries Australia has badly neglected this approach to land management
- demonstration/research is required on shelterbelts for the semi-arid and dry temperate zone — with emphasis on belt width and species composition in relation to their effectiveness in protecting fields; optimal spacing of belts across paddocks; overcoming competition with crops along the belt margins
- in all regions more data is needed on the relationship between shelter and agricultural response — particularly for crops — models such as 'Farmtree' [23] currently do not have an adequate base of data for tree and plant interactions
- in those pastoral regions where there is presently woodland there is an

urgent need to define the long-term effects of clearing on sustainable landuse — the present information is simply not adequate — and to develop alternatives to tree removal in order to attain or maintain a sustainable production system

— agroforestry options (woodlots and timberbelts) need to be investigated for all regions but particularly for the dryland areas where grazing and/or cropping activities have been ecologically destructive.

References

1. Anderson GW (1989) Agroforestry for Productivity. Proc third agroforestry conference, Morwell, Victoria. Victorian Agroforestry Extension Committee
2. Bagnold RA (1941) The Physics of Blown Sand and Desert Dunes. Chapman and Hall, London
3. Bates CG (1911) Windbreaks: Their Influence and Value. US Dept of Agric, Forest Services Bulletin 86. Govt Printing Office, Washington
4. Beckwith RS (1987) Low solubility phosphate fertilisers. Agric Sci 1: 18—26
5. Bicknell D (1991) The role of trees in providing shelter and controlling erosion in the dry temperate and semi-arid southern agricultural areas of Western Australia. In: Proc of a national conference at Albury: The Role of Trees in Sustainable Agriculture, pp 21—39. Publ National Agroforestry Working Group in conjunction with the Bureau of Rural Resources, Dept Primary Industries and Energy, Canberra
6. Bird PR (1984) Effects of trees on agricultural productivity. In: Proc National Conf Focus on Farm Trees 2, pp 13—29. Univ of New England, Armidale, NSW, Canberra Publ & Printing Co, Fyshwick, ACT
7. Bird PR, Lynch JJ and Obst JM (1984) Effect of shelter on plant and animal production. Proc Aust Soc Anim Prod 15: 270—273
8. Bird PR (1988) Timber production from shelterbelts. In: Proc seminar: Tree Growing on the Basalt Plains. Dept Cons, Forests & Lands/Rural Trees Aust, Streatham, Victoria
9. Bird PR (1988) Financial gains of trees on farms through shelter. In: Proc International Forestry Conference for the Australian Bicentenary, Vol 2. Aust Forest Develop Institute, Albury-Wodonga
10. Bird R (1991) Tree and shelter effects on agricultural production in southern Australia. Agric Sci 4(5): 37—39
11. Bird PR (1991) The role of trees in protecting soils, plants and animals in Victoria. In: Proc of a national conference at Albury: The Role of Trees in Sustainable Agriculture, pp 77—87. Publ National Agroforestry Working Group in conjunction with the Bureau of Rural Resources, Dept Primary Industries and Energy, Canberra
12. Bird PR and Cayley, JWD (1991) Bad weather, shelter and stock losses. Agric Sci 4(4): 18—19
13. Brandle JR, Hintz DL and Sturrock JW (1988) Windbreak technology — Proc of International Symp at Lincoln, Nebraska. Elsevier, Amsterdam
14. Bulman PA (1991) Shade, shelter and erosion control — a simplistic model to compare the relative benefits of different farm tree designs for South Australia. In: Proc of a national conference at Albury: The Role of Trees in Sustainable Agriculture, pp 41—50. Publ National Agroforestry Working Group in conjunction with the Bureau of Rural Resources, Dept Primary Industries and Energy, Canberra
15. Burke S (1991) Effect of shelterbelts on crop yields at Rutherglen, Victoria. In: Proc of a national conference at Albury: The Role of Trees in Sustainable Agriculture, pp 89—99. Publ National Agroforestry Working Group in conjunction with the Bureau of Rural Resources, Dept Primary Industries and Energy, Canberra

16. Caborn JM (1957) Shelterbelts and Microclimate. Forestry Commission Bull No 29, Edinburgh, UK
17. Cameron DM, Rance SJ, Jones RM, Charles-Edwards DA and Barnes A (1989) Project STAG: an experimental study in agroforestry. Aust J Agric Res 40: 699—714
18. Campbell A (1991) Planning for Sustainable Farming — the Potter Farmland Plan story. Melbourne, Victoria, Publ Lothian Books, Greening Aust
19. Campbell CA (1990) Planning. In: Cremer KW, ed, Trees for Rural Australia, pp 307—317. CSIRO, Inkata Press, Melbourne and Sydney
20. Carter DJ and Humphrey MG (1983) The costs of land degradation. Journal of Agriculture, Western Australia 24: 50—53
21. Chapman M (1990) Amenity. In: Cremer KW, ed, Trees for Rural Australia, pp 336—355. CSIRO, Inkata Press, Melbourne and Sydney
22. Commonwealth of Australia (1979) A Basis for Soil Conservation Policy in Australia. Report No. 1, Australian Govt Publ Service, Canberra, 193 pp
23. Daly JJ (1984) Cattle need shade trees. Qld Agric J 110(1): 21
24. Davidson RL and Briggs GM (1988) Farm Trees: A Practical Guide. Publ Focus on Farm Trees 2 Committee, Univ New England, Armidale, NSW
25. Donnelly JR (1984) The productivity of breeding ewes grazing on lucerne or grass and clover pastures on the tablelands of southern Australia. III Lamb mortality and weaning percentage. Aust J Agric Res 35: 709—721
26. Eckersley R (1989) Re-greening Australia: the environmental, economic and social benefit of reforestation. Occasional paper No. 3, CSIRO
27. Ecos (1980) Vol 25. Soil Erosion: Can We Dam the Flood?
28. Farmtree — a Computer Model to Help Estimate the Profitability of Agroforestry (undated). Dept Agriculture/Dept of Conservation and Environment, Melbourne, Victoria
29. Geiger R (1965) The Climate Near the Ground. Harvard Univ Press, Cambridge, Massachusetts, 507 pp
30. Hagan LJ (1976) Windbreak design for optimum wind erosion control. In: Proceedings of the symposium: Shelterbelts on the Great Plains, pp 20—22. Denver, CO
31. Hagan LJ and Skidmore EL (1971) Turbulent velocity fluctuations and vertical flow as affected by windbreak porosity. Trans of the ASAE 14: 634—37
32. Hall N, Boden RW, Christian CS, Condon RW, Dale FA, Hart AJ, Leigh JH, Marshall JK, McArthur AG, Russell V and Turnbull JW (1972) The use of trees and shrubs in the dry country of Australia. Aust Govt Publ Service, Canberra
33. Helyar K (1991) Soil acidity — causes, concerns and challenges. In: Proc seminar: Everlasting Plants and Soil — Is It Possible. Western District branch of Grassland Soc of Vic, Hamilton, Victoria
34. Houghton D (1984) Trees and erosion control. Qld Agric J 110(1): 9—12
35. Hughes KK (1984) Trees and salinity. Qld Agric J 110(1): 13—14
36. Institute of Foresters of Australia (1989) Trees: Their Key Role in Rural Land Management. Submission to the House of Representatives Committee of Inquiry into Land Degradation in Australia
37. Irvin R (1991) Portugal's beleaguered Cork Oak. Ceres No. 127, 23(1): 37—42
38. Knight M (1989) Preliminary shelterbelt wind-test results in mature random spaced open woodlands. In: Proc seminar: Land Care Futures, Melbourne. Landscape Architecture Unit, Royal Melbourne Institute of Technology
39. Kort J and Brandle JR (1991) WBECON: a windbreak evaluation model. 1. A Comparison of windbreak characteristics. In: Proc third international symposium: Windbreaks and Agroforestry, pp 129—131. Ridgetown College, Ontario, Canada
40. Kort J and Brandle JR (1991) WBECON: a windbreak evaluation model. 2. Economic returns from a windbreak investment in the Great Plains. In: Proc third international symposium: Windbreaks and Agroforestry, pp 131—134. Ridgetown College, Ontario, Canada

41. Kulik KN (1991) Aerial and satellite methods of research in agrosilvicultural amelioration. In: Proc third international symposium: Windbreaks and Agroforestry, pp 55—59. Ridgetown College, Ontario, Canada
42. Leys J (1991) The role of trees in providing shelter to crops and controlling erosion. In: Proc of a national conference at Albury: The Role of Trees in Sustainable Agriculture, pp 63—75. Publ National Agroforestry Working Group in conjunction with the Bureau of Rural Resources, Dept Primary Industries and Energy, Canberra
43. Lipsett J and Dann PR (1983) Wheat: Australia's hidden mineral export. J Aust Inst Agric Sci 49(2): 81—89
44. Loveday J (1988) Australian soils — the last 200 years. In: Review Papers, National Soils Conference, Canberra
45. MacLennan H (undated) Cultivation of arable land. Soil Conservation Authority, Soils Aid No 23. Govt Printer, Melbourne, Victoria
46. Marsh B and Carter D (1983) Wind erosion. J Agric, WA 24(2): 54—57
47. Marshall CJ (1990) Control of erosion. In: Cremer, KW, ed, Trees for Rural Australia, pp 369—376. CSIRO, Inkata press, Melbourne and Sydney
48. Murray JS and Mitchell A (1962) Red gums and the nutrient balance. Proc Third Aust Conf Soil Science, Canberra
49. Parker JN (1991) The role of trees in shade, shelter and control of erosion in Tasmanian agriculture. In: Proc of a national conference at Albury: The Role of Trees in Sustainable Agriculture, pp 101—107. Publ National Agroforestry Working Group in conjunction with the Bureau of Rural Resources, Dept Primary Industries and Energy, Canberra
50. Pearson HA and Lewis CE (1989) Agroforestry in the South-Eastern United States, pp 1637—8. XVI International Grassland Congress Nice, France
51. Reid R and Bird R (1990) Shade and shelter. In: Cremer KW, ed, Trees for Rural Australia, pp 319—335. CSIRO, Inkata Press, Melbourne and Sydney
52. Reid R and Wilson G (1985) Agroforestry in Australia and New Zealand. Publ Goddard and Dobson, Victoria, Australia
53. Rural Research (1989) No 144. Wind erosion — dust, dunes, and tell-tale sediments, pp 13—23
54. Rural Research (1989) No 144. Tree clearing in the semi-arid tropics, pp 24—28
55. Select Committee into Land Conservation (1990) Discussion paper No 2: Agricultural region of Western Australia. Legislative Assembly, Western Australian Parliament
56. Stepanov AM and Malanina ZI (1991) Shelterbelts in the USSR: an integral part of the agrolandscape. In: Proc third international symposium: Windbreaks and Agroforestry, pp 52—55. Ridgetown College, Ontario, Canada
57. Sturrock JW (1969) Aerodynamic studies of shelterbelts in New Zealand. 1. Low to medium height shelterbelts in mid Canterbury. NZ J Sci 12: 754—76
58. Sturrock JW (1972) Aerodynamic studies of shelterbelts in New Zealand. 2. Medium height to tall shelterbelts in mid Canterbury. NZ J Sci 15: 113—40
59. Song Zhaomin (1991) A review of the development of shelterbelt system in China. In: Proc third international symposium: Windbreaks and Agroforestry, pp 139—143. Ridgetown College, Ontario, Canada
60. The Victorian draft Decade of Landcare plan — towards sustainable land use (1991) A discussion paper prepared for public comment by the Decade of landcare plan steering committee. Dept of Premier & Cabinet, Victoria
61. USDA Forest Service (undated) The Center for Semiarid Agroforestry. Rocky Mountain Forest & Range Experiment Station, Forestry Sciences Laboratory, Lincoln, Nebraska
62. van der Sommen FJ (1991) Amenity plantings in the Northern Territory. In: Proc of a national conference at Albury: The Role of Trees in Sustainable Agriculture, pp 51—54. Publ National Agroforestry Working Group in conjunction with the Bureau of Rural Resources, Dept Primary Industries and Energy, Canberra
63. Voller P (1991) The role of trees in providing shelter and controlling erosion in Queens-

land. In: Proc of a national conference at Albury: The Role of Trees in Sustainable Agriculture, pp 55—61. Publ National Agroforestry Working Group in conjunction with the Bureau of Rural Resources, Dept Primary Industries and Energy, Canberra

64. Water Authority of Western Australia (1989) Stream Salinity and Its Reclamation — South-West Western Australia. Report No WS 52, Report of steering committee for research on land use and water supply
65. Western Australian Department of Agriculture and Water Authority of Western Australia (1988) Situation Statement: Soil and Land Conservation Programme in Western Australia. Submission to the House of Representatives Enquiry into Land Degradation
66. Williams P (1989) In: Land and Water Research News No 4: 3—4
67. Williams J (1991) Search for sustainability: agriculture and its place in the natural ecosystem. Agric Sci 4(2): 32—39
68. Wilson JR (1990) The eleventh hypothesis: shade. Agroforestry Today 2(1): 14—15
69. Wilson G, Kellas J and Kirby G (undated) Livestock Havens. Dept of Conservation and Environment/Dept of Agriculture, Victoria
70. Wood O (1985) Crops, trees thrive at Rich Avon. Stock and Land 30 May, p 31
71. Young A (1989) Ten hypotheses for soil agroforestry research. Agroforestry Today 1(1): 13—16
72. Xiang Kai-fu (1991) Shelterbelt systems and agroforestry in northern China. In: Proc third international symposium: Windbreaks and Agroforestry, pp 9—12. Ridgetown College, Ontario, Canada

Agroforestry Systems **20**: 87–115, 1992.

The role of trees in sustainable agriculture – an overview

R. T. PRINSLEY
Bureau of Rural Resources, DPIE. Present Address: Rural Industries Research & Development Corporation, PO Box 4776, Kingston, ACT 2604, Australia

Key words: agroforestry, Australia, soil windbreak, timber, farm

Abstract. Agroforestry in the Australian environment has potentials to rehabilitate land from further degradation; improve agricultural productivity; produce timber; *and* contribute to diversification and increase in farm income. These benefits are examined using available information from research. The level of adoption of agroforestry by Australian farmers for these purposes is also determined. Summaries of research results provide the basis for advice that scientists can offer to farmer and decision makers.

1. Introduction

Agriculture is a major component of Australian exports contributing 24.4% of exports in 1989/90 [ABARE, 1990]. Sustainable agriculture is dependent upon the continued availability of suitable land and water. Alternative approaches to land management in Australia must be developed and supported and these must have objectives of maintaining long-term as well as short-term productivity of agricultural land while minimising water pollution. Evidence of the concern for the environment in the community can be seen locally, at the state level and at the national level (Australia is a federal nation comprising six states and two territories). This concern includes the development and action at 'grass-roots' level of hundreds of community groups such as 'landcare' and farm-tree groups. At the state level there are many schemes which assist landowners and community groups to rehabilitate and conserve our natural resources: these include a large number of tree-planting assistance schemes [Boutland et al., 1990]

The Federal Government has instigated and sponsored programs which support sustainable land management and conservation. A number of these relate to agroforestry. The Ecologically Sustainable Development Working Groups on Agriculture and Forestry have formulated strategies for sustainable development and are providing advice to the Government on practical proposals for implementing them. Both refer to various benefits of agroforestry and maintaining native vegetation on farms. The 1990's have been designated the Decade of Landcare, and the National Soil Conservation Program is supporting hundreds of landcare groups to practise sustainable agriculture. The One Billion Trees program is dedicated to establishing one billion trees in Australia by the year 2000, while the Save the Bush program is dedicated to preserving and managing native vegetation.

At the same time, fluctuations in farm income combined with the continuing cost-price squeeze represent major problems for Australian farmers [Ferguson, 1988]. Diversification into products from trees may provide some insurance against such uncertainties in market reliability for more traditional agricultural products.

The integration of trees into farming systems — agroforestry — has the potential to improve both sustainability and profitability of Australian farms. Trees from farms can also provide a range of timber products in substantial quantities for the Australian timber industry, thus reducing our import bill for wood products. At present it is estimated that 26% of farmers in Australia are establishing trees for various purposes. Most of these are in the states of Victoria and Western Australia, where 50—60% of farmers plant trees [Prinsley, 1991a]. In NSW and Queensland, however, less than 10% of farmers appear to be involved in agroforestry.

There are many purposes for establishing trees on farms or protecting native woody vegetation, including salinity control, windbreaks, shade and erosion control, conservation of biodiversity, and production of timber, fodder and other products e.g. oils, nuts and wildflowers. In many cases, the same tree can be used for more than one purpose.

The papers in this edition of *Agroforestry Systems* are from major national reviews prepared for a national conference on 'The Role of Trees in Sustainable Agriculture' held in October 1991. They provide information on the purposes for agroforestry in Australia and also on the latest research on agroforestry in Australia.

Statements are made repeatedly about the benefits of agroforestry; but these are not always successful in convincing farmers or decision makers that trees have a role in agriculture. It is suggested that trees have the potential to achieve simultaneously one or more of the following categories of benefits:

— rehabilitate land from further degradation;
— improve productivity e.g. via windbreaks;
— produce timber;
— contribute to diversification and increase in farm income.

Information provided by national review papers in this edition is evidence on which to examine these statements. Each will be discussed in this paper, remembering that in many cases, benefits are derived from practising agroforestry for more than one purpose.

Each section will consider opportunities for agroforestry and problems which can be addressed in different parts of Australia. Where possible, it will be determined how commonly these opportunities are taken up. Each chapter will also summarise past and present contributions of research. This can form the basis for the advice that scientists can now offer to farmers and decision makers. Examples of promising and innovative research will be described.

The purposes and practices of agroforestry may vary depending upon a combination of climate, soil, topography, land use, socio-economic and policy factors. It is usually impossible to recommend a specific practice for the whole of Australia. It is often feasible, however, to consider whether an agroforestry practice will suit a particular agroclimatic zone (Fig. 1) within the categories shown.

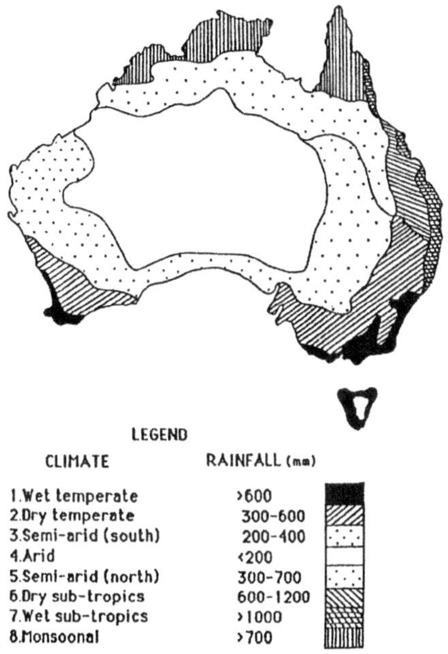

Fig. 1. Agroclimatic zones of Australia (Source: National Farmers Federation 1991).

Each state and territory has its own government and related policies. Thus in some cases the State or Territory in which the farm lies may have a significant effect upon the policy and socio-economic climate which can determine the type of agroforestry practice implemented. For instance, in Western Australia, due to government and industrial incentive schemes, it is currently economically attractive to plant hardwoods for commercial pulp production and large numbers of farmers have responded. In contrast, in Queensland, where the Forest Service has established extensive areas of pine plantations, there is very little incentive for commercial agroforestry and very little activity even though the natural environment is suitable for agroforestry. Therefore, throughout this paper and the other papers in this conference series, specific states, land uses and agroclimatic zones will be referred to where necessary.

2. Agroforestry for land rehabilitation and protection

Since Europeans settled in Australia in the late 18th century, there has been clearing of vast areas of forest and woodland to expand agricultural production. This has had, and is still having, many direct and indirect harmful effects on the environment. Figures 2a and 2b show the reported extent of these effects and elicit the following questions. What evidence is there that replacing trees will restore the environment and protect it from further damage? How can trees be integrated with farming to protect the environment and maintain productivity?

Damage to the environment can be on a local (farm) or on a broader (catchment) scale. The environmental benefits of hardwood plantations and agroforestry on cleared agricultural land are reviewed in some detail by Cadman, Prinsley and Moxham [1991]. Their review shows that tree planting may:

- reduce salinity
- improve soil fertility
- control and prevent erosion
- control waterlogging
- reduce catchment eutrophication
- possibly check acidification
- probably increase local biodiversity
- possibly reduce the Greenhouse Effect

The major foci of the environmental problems caused directly or indirectly by tree removal are shown in Fig. 2c. Most of these environmental problems, which can be ameliorated by trees, occur widely in Australia. Of these effects, this paper will address the ways in which agroforestry can control problems which currently most directly affect agriculture. These include salinity and soil erosion, and improvement of soil fertility. Most research in Australia is concentrating on the role of trees in controlling salinity [Prinsley, 1991a]. Nick Schofield (review in this issue) has prepared a national review of salinity research in Australia which is summarised and discussed in the following section on dryland salinity. Schofield emphasises the complexity inherent in the integration of agroforestry research results with farming systems.

2.1. *Dryland salinity*

2.1.1. *Cause*

The major cause of dryland salinity has been the clearing of deep rooted perennial native forests and woodlands and their replacement with shallow rooted annual agricultural species. The reduced consumption of water by annual crops and pastures leads to increased groundwater recharge and rising groundwater levels. Where there is a source of salts in the soil profile,

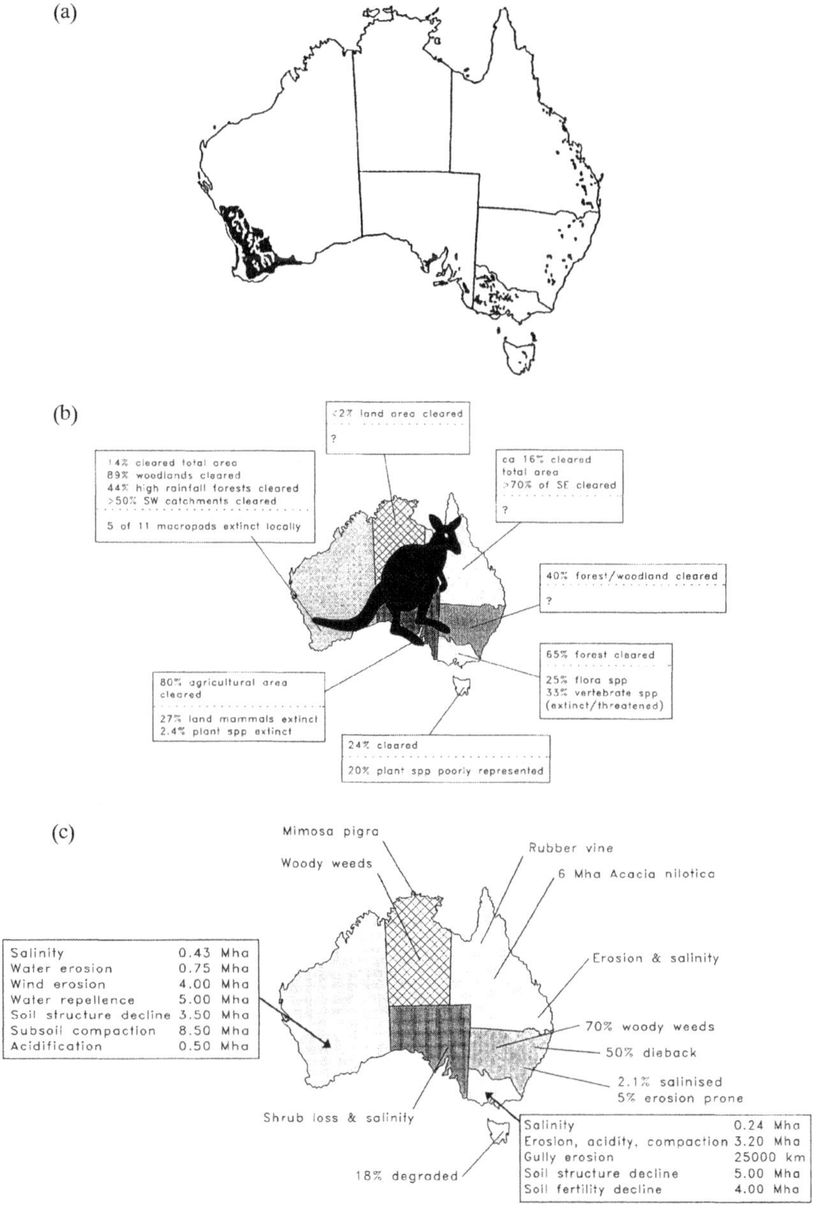

Fig. 2. (a) Areas susceptible or containing dryland salinity in Australia (figure should be considered schematic as definitions of dryland salinity vary from State to State) (Copied from Schofield, this issue); (b) Clearing and loss of threatened species in each State (Information collated from the reviews in this issue); (c) Land degradation in States and Territories of Australia (Information collated from the reviews in this edition)

rising groundwater brings the salts to the surface. Salinity usually decreases the productive potential of the land, decreases the water resources available for consumption, and degrades the ecological values of wetlands and rivers and their associated vegetation and habitats [Schofield, this issue].

2.1.2. *Extent and treatment*

The extent of dryland salinity is indicated in Fig. 2a. It has been identified as a major form of land and water degradation in southern Australia, particularly in Western Australia, South Australia and Victoria, and to a lesser extent in New South Wales, Tasmania and Queensland. Tree planting, in combination with other vegetation treatments, is regarded as a leading solution to dryland salinity [Schofield, this issue].

2.1.3. *Criteria for selection of species for salinity control*

Three important criteria are often quoted when addressing tree selection:

- The trees must be adapted to site conditions.
- Their potential water use should be predictable for that site.
- The trees should have uses, such as timber, or shelter for crops or animals, as well as salinity control.

Research in both Western Australia and Victoria has been carried out to select salt tolerant species and provenances which can be planted in discharge areas. Some of the most tolerant have been targeted to cover a range of climatic conditions throughout Australia and to serve various economic uses [McComb et al., 1991].

2.1.4. *Where to plant trees and how many*

Schofield points out that it is difficult to determine where and at what density to plant for salinity control because of the number of configurations possible. This problem has been the target of substantial modelling work referenced by Schofield (this issue). This work starts with the requirement that annual evapotranspiration plus streamflow from the land must equal or exceed the rainfall to reduce accession to groundwater. Also farming practicalities influence where and at what density trees can be planted.

2.1.5. *Applying salinity research to farms*

About 16% of farmers establishing trees in Australia are doing so for salinity control. Of these, 7% are to prevent salinity, while 9% are to reclaim saline land [Prinsley, 1991]. In many cases their successes and failures are known locally but the information is not publicised. Thus it is difficult to know how successful their efforts are, or what strategies they are using.

The theory of tree planting for salinity control is largely understood. Theory is constrained, however, by a number of practicalities. Thus salinity research is a good example of the difficulties of applying research results to farms. It demonstrates the need for on-farm research and research carried out in consultation with farmers from the outset, so that both farmers' and technical requirements are met.

The needs of the farmer may dictate that choice from among several well-adapted species should be based on economic criteria. If there are direct benefits from the trees such as timber and the water use potential is quite high, then a particular species can gain preference. Optimising the commercial potential of tree planting may then govern such things as planting density and location. Choosing species on the basis of water use alone is difficult for many sites because of the lack of comprehensive, good quality data.

A second constraint is agricultural manageability. Whereas woodlots or wide-spaced trees may potentially be more efficient water users, they are also obstacles to many farming operations. As a consequence there has been a move towards tree belts which allow cropping and other farm practices to occur unhindered between the belts. Whilst this system has clear practical advantages, it poses questions in regard to optimum belt width, belt spacing and planting densities within belts for preventing the rise of water tables. Recharge may also occur between belts. An alternative to systematic schemes such as wide-spaced or belt plantings is high recharge area planting. Considerable study has been devoted to identifying potentially high recharge areas [Walker et al., 1991].

The landowner's perceptions and desires can also strongly influence species selected, where trees are planted, and how many.

Although numerous examples of lowering groundwater levels by trees have now been reported, there has been less data published on the consequent effects on groundwater and soil salinity. Research has indicated, however, that salt dynamics in relation to tree planting requires further careful research in lower rainfall (and irrigation) areas where salt levels might increase and kill the trees. (This may not be necessary in higher rainfall areas.)

2.1.6. *Salinity research needs*

Salinity research for agroforestry is one of the more advanced areas of agroforestry research in terms of the numbers of research projects [Prinsley, 1991a] and the advances made to date. Sufficient data are available to demonstrate the potential of tree planting as a salinity control measure [Schofield, review in these proceedings]. In specific environments it is possible to determine appropriate species or provenances to plant, how and where to plant them, and in what densities and configurations. However, these site-specific strategies cannot yet be applied successfully to a wide range of sites.

In contrast, in an established industry such as wheat, hundreds of different varieties of wheat have been bred, providing optimum characteristics for a wide range of conditions. The management and treatment requirements of each variety are well understood. Crop growth models and decision support systems based upon these models are available to assist the farmer or extension officer to select the most productive or least risky variety and management regime for a chosen environment. These models have been only recently developed by collation of much experimental data. To realise the potential for trees to control salinity, more research needs to be done to bring our understanding and ability to implement it to the same level as wheat production. Questions of high priority include:

- Advances in tree technology: selection, cloning, breeding hybridisation, productivity and site amelioration for improved establishment will provide further gains.
- Modelling can make a major contribution. Strategic surveys (electromagnetic, soil, geology, topography etc.) combined with modelling will increasingly become a necessary preliminary stage to developing a vegetation strategy for salinity control. Field trials will continue to be necessary for validation of models and as demonstrations. Unlike wheat, where trials were done all over Australia in an unsystematic manner and information was compiled much later for model development, there is the opportunity in agroforestry to carry out trials as verification of models, thus maximising the efficiency and use of limited research resources.
- In some hydrogeologic systems, there is a need for more study of possible salt accumulation beneath trees, and its effects [Schofield, this issue].

2.2. *Agroforestry for soil conservation*

The beneficial effects of trees on agricultural systems is a recent area of investigation. Thus, unlike other, more traditional agricultural industries, such as wheat, with a huge body of knowledge behind them based on many thousands of research trials, agroforestry has only a small amount of data from research in limited agroecological zones on which to base recommendations to farmers. Trees, as compared to agricultural crops, often incur a further research disadvantage of a long growing period, so that in many cases it can take a working lifetime or more to finish an experiment.

Many 'farm-notes' produced by government and advisory agencies concerning agroforestry often contain recommendations based upon very little or no quantitative data, or on data obtained from a different agroecological zone. They may consist of unquantified general statements which do not inspire confidence in the farmer. However, there is often no alternative because the appropriate data are not available. Most of the stated capabilities of agroforestry are still at the hypothesis stage or the 'demonstrated for some sites' stage. Very few are at the stage of research where they are quantified technically and economically, and able to be applied confidently in all areas where they are needed. The following section on agroforestry for soil conservation (excluding salinity) will attempt to classify the information provided on the basis of its stage of research, and whether research results can be reliably extended to a range of sites.

In Australia between a half and two thirds of the land used for grazing or cropping (>2.6 million km^2) needs treatment for land degradation [Eckersley, 1989]. Figure 2a shows the extent of different forms of land degradation in Australia as reported by the review papers at this conference. It has been estimated that for each dollar that dryland salinity costs us, water and wind erosion costs $5, soil acidification $25, soil structural decline $125 and soil nutrient degradation $625 [Institute of Foresters of Australia, 1989]. It is clear from these figures that greater emphasis should be placed on controlling soil erosion and associated nutrient loss and improving soil fertility.

2.2.1. *The processes of soil amelioration by trees*

Tree replacement will be essential if we are to re-develop the fertility of many of our degraded soils and maintain sustainable systems of agriculture. In fact, of the farmers establishing and protecting trees in Australia, about 13% are estimated to do so for soil conservation [Prinsley, 1991a]. This does not include another 16% of farmers establishing trees, who are planting them for salinity control [Field, 1989].

Ingram [1990] and Young [1989] have reviewed the role of trees in maintaining and improving soil productivity. Table 1 lists the processes by which trees improve soils. Bird et al. (this issue) also discuss the role of trees in protecting soils. Information provided by Young, Bird et al. and Ingram is used to denote the stage of research concerning each listed process.

These processes are discussed in further detail by Ingram [1990], Young [1989] and Bird et al. in this issue. There is considerable scope for further research, in particular, concerning all the processes still at the hypothesis or demonstration stage. Despite huge losses to productivity in Australia due to soil degradation [Institute of Foresters of Australia, 1989], most of the research has been done outside Australia except for salinity research. There is thus great potential for some of the successful practices demonstrated overseas to be adapted through research to Australian conditions. It is also interesting to consider why tree planting for soil conservation other than for salinity is given such a low priority both in implementation and in research. In other countries where productivity is reduced by soil degradation, the use of agroforestry in strategies to ameliorate these losses is being investigated more thoroughly.

2.2.2. *Erosion*

Clearance of native vegetation for agricultural land use has been the major cause of accelerated erosion in historical times in Australia. The largest pulses of soil movement and loss generally occur within the first few years after clearing [Ingram 1990]. However, continued soil erosion, by both water and wind, occurs on all exposed land surfaces as a result of cropping and animal grazing practices. There has been very little research carried out on the role of trees in preventing soil erosion. Currently there are four research

Table 1. Processes by which trees improve soils (adapted from Young 1989).

Increasing inputs
- organic matter by photosynthesis (**P**)
- nitrogen fixation (**P**)
- nutrient uptake and recycling from deeper soil horizons (**H**)

Reducing losses
- by protection from erosion (**D**)
- of nutrients otherwise leached by promoting recycling (**D**)
- of organic matter (by shading) (**D**)

Improving soil physical conditions
- soil structure (**D**)
- porosity (**D**)
- moisture characteristics, water-holding capacity (**D**)
- erosion resistance (**D**)
- breaking up compact layers by roots (**D**)
- modification of extremes of soil temperature (**D**)

Improving soil chemical conditions
- reduction of acidity (**H**)
- reduction of salinity (**D**)

Beneficial effects on soil biological processes
- production of a range of qualities of plant litter (**D**)
- timing of nutrient release (**H**)
- effects upon soil fauna (**H—D**)
- transfer of assimilate between root systems (**H**)

The letter, in brackets, is assigned to each process and refers to the following:
H — hypothesis
D — demonstrated and widely accepted; we know how to use it in specific circumstances only
P — proven and quantified. Usually possible to apply to a wide range of conditions

projects Australia-wide in this area, compared with 42 research projects on salinity research [Prinsley, 1991a].

2.2.2.1. *Water erosion.* Water erosion accounts for about 70% of the total land degradation Australia-wide [Commonwealth of Australia, 1979]. Tree clearing for agriculture and soil disturbance associated with subsequent grazing and cultivation has produced massive soil erosion. For instance, in the wheatlands of Queensland the soil loss may exceed 50 t/ha/y (i.e. around 4 mm/y), which is around 50 times the most optimistic estimate of the rate of soil formation [Ecos, 1980]. Soil erosion has a direct local effect on the productivity of the site and an indirect effect by increasing sediment deposition in reservoirs.

The cover provided by tree litter is the major way in which trees prevent soil erosion. Species with slower leaf decay such as eucalypts may provide longer lived litter cover of the soil [Young, 1989]. Litter accumulation at the surface can also impede water flow, giving it greater time to be absorbed by the soil. The movement of water deeper into the soil can be influenced by the presence of open paths created by roots and associated soil fauna and flora.

Trees have a large role in controlling soil erosion on poorer classes of agricultural land such as steep areas, degraded or eroded land and land with soil structure or fertility problems; here removal of trees is detrimental because of the difficulty in establishing crops or pastures after clearing. In many other cases trees alone cannot protect the soil from water erosion, but integration of trees with agriculture to provide an extra ground cover of litter can provide the best protection [Marshall, 1990].

Young [1989] has described the various ways in which trees can be usefully integrated into agroforestry to control erosion. He states that, with the exception of extreme cases, the major adverse effect of erosion is lowering of crop yields through loss of organic matter and nutrients in sediment and run-off. Young suggests that there is a substantial potential for reducing erosion through the use of agroforestry-based methods and summarises the following functions of trees and shrubs in erosion control under an agroforestry situation:

Direct use

- to increase soil cover, by litter and prunings
- to increase soil resistance to erosion, by maintenance of organic matter

Supplementary use

- to stabilise earth structures by root systems
- to make productive use of land occupied by conservation works

2.2.2.2. *Wind erosion.* Wind erosion of agricultural or pastoral lands is a severe problem world-wide. Highly productive dryland agricultural ecosystems in Australia are very vulnerable to wind erosion. Current approaches in North America and the USSR described by Bird et al. (this issue) indicate how they are dealing with soil erosion through tree establishment. For instance, in the USSR, about 40 million ha of the 90 million ha of cropped land subject to wind erosion are protected by 2 million ha of shelterbelts; the combined effects of improved soil cultivation technology and shelterbelts have reduced average soil losses to around 20% of previous rates, with even better results in some regions. Wind erosion is substantial across the Australian continent and particularly in the semi-arid lands in south western New South Wales, South Australia, Victoria and Western Australia [Rural Research, 1989].

Bird et al. (this issue) discuss strategies for control of wind erosion. 'Minimum tillage' and stubble management in cropping have been a major extension and research activity of the Western Australian Department of

Agriculture for nearly a decade. This reliance on management of crops and annual pastures for wind erosion control could be adequate, if universally practised, in a 'normal' year. However, severe weather and/or economic factors, combined with less than perfect adoption of appropriate grazing and crop management systems, show the weakness of relying completely on annual management decisions. Tree windbreaks for prevention of wind erosion can help to provide an insurance against severe conditions with which normal good management practices cannot cope. Combinations of tree and agronomic protection systems are likely to give higher returns over a range of risk conditions. Research in Victoria by Bird et al. (this issue) has confirmed very extensive overseas research [Geiger, 1965; Bates, 1911; Caborn, 1957; Sturrock, 1969; Sturrock, 1972] that a variety of shelterbelts afford significant protection from wind.

In conclusion, wind erosion in some parts of Australia is a major problem (see Fig. 2c), and we have information that allows us to reduce windspeed by 50% using windbreaks. An estimated 17% of farmers in Australia who practise agroforestry are establishing trees for windbreaks [Prinsley, 1991a]. Nevertheless, compared to other countries, Australia has neglected this approach to land management. Hence research into the effects of windbreaks of various designs in combating soil erosion should receive high priority, particularly in the arid, semi-arid and dry temperate zones [Bird et al., this issue].

3. Agroforestry for improving the productivity of agriculture

3.1. *Introduction*

Agroforestry has a major role in controlling land degradation and in maintaining soil fertility. These are measures which maintain the condition of soil and water, thus indirectly increasing agricultural productivity. However trees can have other roles, apart from those described so far, for directly improving agricultural productivity in the shorter term. These include shelter for crops, pastures and livestock which can be provided both by planted trees and by native vegetation.

3.2. *Shelter*

3.2.1. *Shelter benefits for crop and pasture production*

It has been estimated that 14% of farmers who practise agroforestry are establishing and protecting trees for shade, another 14% for shelter, and about 17% for windbreaks [Prinsley, 1991a]. Thus almost half of the farmers in Australia practising agroforestry are doing so for shelter benefits. These farmers perceive that there are advantages in planting trees for shelter. But what are the benefits of shelter for crop and pasture production? Bird et al.

(this issue) include the following findings. (notations after the finding: H = hypothesis, D = demonstrated and P = proven and quantified):

- Reduced water loss by shading and/or reducing windspeed
  - this can also prolong pasture growth in summer [Reid and Bird, 1990; Bird et al., 1984] (**D**).
- Temperature modulation
  - by reducing inward radiation during the day and decreasing outward radiation at night [Geiger, 1965](**D**).
- Reduced mechanical damage and sandblasting
- Control of disease and insects
  - windbreaks may assist in the control of anthracnose outbreaks in mangoes [A. Mead, pers. comm.]. Windbreak trees may also afford habitat for predatory birds and insects which may help control some insect pests (**H**).
- Improved pollination efficiency
  - e.g. fruit set of cucurbit crops in Queensland may be increased 30% in some years [R. Goebels, pers. comm.] (**H-D**).
- Control of spray drift
  - shelterbelts are used to reduce the drift of chemical sprays from cotton fields in Queensland; the windbreaks also extend the opportunities for spraying in otherwise unsuitable weather conditions (**H-D**).
- Improved production from orchard fruit
  - in Tasmania some stone-fruits can be very responsive to shelter [S. Watson, pers. comm.] (**H**).
- Direct improvement of crop yield
  - Shelter provided by eucalypts can increase wheat and oat crop yields in the sheltered zone by 25% and 45%, respectively [Burke, 1991]. These data confirm experimental results world-wide [Bates, 1911; Bird, 1984; Bird, 1988; Brandle et al., 1988; Caborn, 1957] (**D**).

3.2.2. *Shelter benefits for animals*

Bird et al. (this issue) also discuss the advantages of shelter to crop, pasture and animal production. The following is a summary of this discussion (the same notations after the finding, H = hypothesis, D = demonstrated and P = proven and quantified are used as in Table 1]:

i) *Temperature modification*
Without shelter, energy expenditure can be increased by excessive heat or cold by diverting energy from productive purposes into maintenance functions.(**D**).

ii) *Fertility enhancement*
Without shelter, heat stress can reduce ram fertility and, particularly with shorn sheep, affect ovulation, oestrus, conception and embryo survival. Heat stressed cows produce smaller calves than those provided with shade, and have longer intervals between calving (**D**).

iii) *Reduced neonatal mortality*
Lamb mortality is decreased by an average of 50% when adequate shelter is provided [Bird et al., 1984]. Since ewes will not always seek shelter prior to lambing, the shelter must either be dispersed or the sheep need to be confined to a sheltered area for maximum effect [Bird and Cayley, 1991; Wilson et al., 1969] (**D**).

iv) *Improved juvenile stock production*
Provision of shade can improve the survival prospects of calves and lambs and subsequent growth and wool production [Bird, 1984; Bird, 1988; Reid and Bird, 1990] (**D**).

v) *Reduced stock losses in new shorn sheep*
Losses of shorn sheep following cold, wet and windy weather in southern Australia can be catastrophic. Shelter protects against such losses [Bird, 1984; Bird, 1988; Reid and Bird, 1990; Bird and Cayley, 1991] (**P**).

3.2.3. *Shelterbelts — the way forward*

The most common reason for establishing trees on Australian farms is for shelter and shade [Prinsley, 1991a]. However, throughout Australia, it was estimated that only 26% of farmers are establishing or protecting trees at all. The major constraint to tree establishment was found to be economic. Economic evaluation of tree planting on farms is crucial to adoption and the few studies which have so far been carried out in this area are thus very important.

Bird et al. (this issue) considered the long term economic effects of establishing a shelter network over a grazing property in rainfed southern Australia. Using a farm financial model they predicted that at discount rates less than about 7% real interest, it was better to have 10% of the farm in a network of close-spaced belts (12.5 heights apart] rather than 5% of the farm in wide-spaced belts. The economics of the system would be further improved by the incorporation of timber species and management of the timberbelts [Bird, 1988]. While this site specific model above gives support to the economic feasibility of a farm revegetation program, a more sophisticated model, such as 'Farmtree' [Loane, 1991] is needed when considering particular species, site and layout for a chosen option.

'Farmtree' is an agroforestry simulation model for micro-computers in which the main features of site, layout, species, intended harvest age, proposed management and other data are entered and likely costs, effects on agriculture, tree growth, timber value and direct effects on farm income and net rate of return are estimated. Extension officers in Victorian regional offices of the Departments of Agriculture and of Conservation and Environment can now use the model to assist farmers and Farm Tree Groups in testing different tree planting options.

A preliminary model for a low rainfall area in South Australia (450 mm) suggests that tree establishment to provide shelter or control salinity would also be profitable in the long term. This model differs from others in that it makes an allowance for land protection values as well as short-term agricultural production factors [Bulman, 1991]. Other models include MULBUD [Etherington, 1991].

Since there is limited data for the environmental inputs into the above models (e.g. the relationship between windbreaks and plant productivity or between soil erosion and plant productivity) these are based upon assumptions and 'best guess' estimates about such relationships. We need to verify the assumptions in the models, particularly in relation to the effects of shelter on plant production (crops and pastures) for a wide range of conditions. It is likely that this evidence will be best defined with the use of empirical and process models validated by field experimentation in different environments.

3.3. *Productivity benefits and costs: managing native vegetation on farms*

This paper has concentrated on the positive effects of trees on agricultural productivity. There are of course, negative effects of trees on agricultural productivity too. If there were not then farmers would not clear trees from their land in order to develop it for agriculture. In the recent past, farmers have had a negative attitude to trees which is still part of the culture of many farmers in Australia. When do trees start to become 'friends and not foes'? This question is being looked at in Queensland. Their systems approach to the problem is possibly the key to successful agroforestry research in Australia [QDPI, 1990] and is discussed below.

Retaining native vegetation on farms in Australia has advantages including direct economic benefits (timber, honey production, shelter for stock), indirect benefits (maintenance of soil fertility), off-site benefits (maintenance of water quality in rivers) and natural heritage value. However, particularly in the tropical states of Australia, Queensland and Northern Territory, where there is a large proportion of native vegetation remaining intact, landholders often see retention of native vegetation as a cost in foregone production. How does the farmer balance the benefits of maintaining trees against clearing them for agricultural productivity? What proportion of trees should be cleared when developing a property?

The latter is the subject of research at the Queensland Department of Primary Industries (QDPI) and is referred to by Scanlan in this edition. However, this research is little known by agroforestry researchers outside Queensland. It is highlighted in some detail here. The outcome of the research carried out by researchers at QDPI is GRASSMAN, which is an easy-to-use computer Decision Support System designed to help beef producers compare strategies for the management of native pastures in eucalypt woodlands in central Queensland [QDPI, 1990]. GRASSMAN is being used by at least one hundred farmers in this area [M. Howden, pers. comm.]. Production of the model was a cooperative effort involving pasture agronomists, woodland ecologists, weed control specialists, modellers, computer programmers and extension agronomists. To develop this understanding, empirical, mechanistic and systems models were formulated:

i) *Tree-grass relationships*
An empirical model of the effects of varying tree basal density on pasture production; this model shows that, underlying the density effect, are a combination of competitive and stimulatory effects. These vary depending upon rainfall, soil type and tree species [Scanlan, 1992].

ii) *Grass production*
An empirico-mechanistic model which demonstrates how grass and consequent livestock production depend upon climate and soil moisture; the effects on grass production of interaction between management strategies, such as stocking rates and burning, are also modelled. It incorporates the above tree-grass model and can determine the competitive effects of trees with grass for water and nitrogen. It also incorporates mechanistic models of water use efficiency.

iii) *GRASSMAN*
This is a 'user-friendly' system model which uses both previous models, as well as economic information concerning beef production and the financial aspects of farm management.

The three models have been verified for eucalypt woodlands with native grass understorey.

Model (i) is the only model in Australia that describes the effects of tree density on pasture production. It shows, for instance, that in eucalypt woodlands, the beneficial effects of trees on the soil (see Section 2) are outweighed by the competition for water and nutrients even at very low tree densities. However, in nutrient poor soils with an overstorey of leguminous shrubs, an individual legume tree has a beneficial effect on pasture, while densities of up to about 300 trees per hectare can also stimulate pasture production. It has been suggested that where clearing of eucalypt woodland has been extensive, useful legume trees such as *Albizzia* or *Leucaena* should be planted in preference to eucalypts, thus permitting stimulated pasture production in some areas as well as fodder production and provision of shade.

The model does not yet incorporate all of the positive effects of trees such as shelter, hydrological and maintenance of biodiversity benefits. In Queensland, strategic clearing and management of the remaining native vegetation can provide improved pasture productivity and also maintain the benefits of trees discussed in this paper. In states and areas where most of the native woodlands and forests have already been cleared for agriculture (Fig. 2), remnant vegetation is often fragmented, environmental problems due to clearing have occurred, and some flora and faunal species are extinct or endangered. In these states and areas, there is often less justification for clearing native vegetation. Figure 2 summarises the information in Scanlan et al.'s paper (this issue) regarding the amount of vegetation cleared in each state, and the effects that this has had on species survival and land degradation. Scanlan et al.'s national review paper (this issue) discusses these issues in some detail, including the legislation and planning regulations associated with clearing.

Scanlan et al. (this issue) examine the situation in Victoria where 65% of forest has already been cleared. In Victoria there appears to be more value in conserving the little remaining native vegetation for environmental reasons rather than clearing it for agriculture. This certainly differs from the situation in Queensland where only 16% of land is cleared and $15000 million worth of beef grown on native pasture is exported each year already. The current research in Queensland concerning native woodland management for grazing, by providing sensible clearing guidelines and vegetation management strategies to farmers, should help to maintain the sustainability of beef production in Queensland.

4. The contribution of agroforestry to the national timber supply

4.1. *Introduction*

Wood growing contributed about $530 million dollars to gross domestic product in 1986—87 (the latest year for which figures are available) while wood processing contributed about $1.2 billion. The timber industry is currently responsible for about 0.5% of total Australian employment and 1% of gross domestic product. Australia produces about two-thirds of the wood and wood products that it consumes. Our major export of these products is eucalypt chips. We imported about $2.1 billion of wood and wood products in 1989—90, leading to a trade deficit of around $1.6 billion for these products, which is equivalent to 7% of Australia's deficit on current account [Resource Assessment Commission, 1991]. There is certainly scope for expansion of timber production for import replacement, particularly of sawn timber, wood-based panels and certain paper products — newsprint, lightweight coated papers and high-strength packaging board. Eucalypt chips and bleached eucalypt kraft pulp offer sound prospects for value-adding through export [Resource Assessment Commission, 1991].

Woodchips of high and even quality from plantations of particular species of eucalypt can command a better export market price than mixed species woodchips from native forest [R. Waining pers. comm.]. Thus there is potential in Australia to develop a coordinated high quality woodchip and pulp industry based upon the establishment of extensive hardwood plantations. This prospect has been one of the subjects of the deliberations of the National Plantations Advisory Committee in 1991. This committee particularly looked at the prospects for a hardwood plantation industry on cleared agricultural land and at how commercial tree crops could be integrated into farming systems and provide environmental benefits.

Most of Australia's land which is suitable for hardwood plantations is either being used for forestry or has been cleared for agriculture [Booth, T. Walker and R.T. Prinsley, unpublished results, 1991]. Farmers are opposed to the state forest services and forestry companies buying farms for plantation establishment in close proximity to processing facilities, and planning restrictions and community opposition prevent farm purchase in many cases [W. McCaughey, C. Humphries, Bunnings, and Department of Conservation and Environment, pers. comm., 1991]. Integration of timber production with farming is not opposed by farmers and has economic potential within reasonable proximity to processing facilities.

4.2. *Incentives and disincentives for timber production from farm*

There are currently various disincentives to the commercial production of timber from farms, not least of which is the high establishment cost at the beginning of the investment and the large number of years it takes to achieve returns which are then taxed highly as a lump sum [Hall, 1990; Hansard and Dean, 1991]. However there are also 28 incentive schemes administered by both industry and government departments to assist farmers to establish hardwood and softwood plantations throughout most of Australia [see Boutland, 1991; Prinsley, 1991a and b]. These schemes usually assist with establishment costs and some of them provide an annuity or lease payment. Many of them are Joint Venture schemes which mean that the farmer and the assistance body share in costs and returns based upon a contract between them. These schemes have been reviewed for their effectiveness for the National Plantations Advisory Committee [Prinsley, 1991b]. Currently while farmers are not experienced in commercial timber production, joint venture schemes provide the farmer with both technical and marketing support. They also provide an assurance to the investor that the required quantity and quality of timber will be produced. In future when farmers have had extensive experience and training, and disincentives to private forestry have been removed there may be more farmers willing to plant trees for timber production independently. Where there are no specific assistance schemes, in general, the type of wood product obtained will be determined by factors such as location and environment, species, management, markets, and proximity to processing facilities.

As discussed earlier, the integration of trees within normal farming practice is seen as a vital element in caring for the land, particularly in the control of soil erosion, nutrient loss and salinity, and the protection of livestock, crops and pastures. Trees also provide a means to diversify farm income where there is a market for the timber products, thus providing a multiple benefit from tree planting [Moore, 1991].

4.3. *Methods of integrating trees on farms*

Methods of integrating trees and farming are being developed. They include plantations or woodlots, timberbelts, windbreaks and wide-spaced stands. In most cases trees are one of several strategies used together to combat land degradation. For example, drainage, perennial pastures and trees can be combined to treat waterlogged and saline areas. Timber production may be secondary to other values, yet it can provide the farmer with an income from the trees planted for these other purposes [Moore, 1991].

4.4. *Types of wood products from farms*

The review by Moore (this issue) has considered the major types of wood products that can be produced on farms in different states and climates in Australia. The review concludes that the integration of trees into farming systems can potentially provide a significant resource to the timber industry, particularly in areas of rainfall higher than 600 mm per annum where there is a processing facility nearby. In drier zones farmers can produce timber for the farm. Currently about 14% of farmers who practise agroforestry in wetter (> 600 mm) rainfall zones of Australia, do so for the purposes of commercial timber production, while a very small number of farmers in drier zones plant trees for sawlog or pulp production [Prinsley, 1991a]. Advances in agroforestry research, removal of disincentives [see Hansard and Dean, 1991] and suitable marketing arrangements, should greatly increase this percentage in the future. Another problem of producing timber on farms is that a single farmer cannot produce an economically viable large volume. However, groups of farmers, such as Farm-Tree or Landcare groups could produce a much larger volume jointly [Prinsley, 1991b].

5. Agroforestry can contribute to an increased and diversified farm income

Economic depression in rural Australia is not uncommon . Vagaries in world prices for commodities, droughts and other climatic events result in a relatively unstable farm income. The establishment of trees can address both climatic and market instabilities.

The previous section discussed how farm timber could contribute to the timber industry. Commercial timber production has the potential to be a major new farm industry in Australia and to thus provide for diversification of farm income. At present this is particularly so where there are assistance schemes for private forestry on farms. For instance, Curly Humphries, wood supply manager for Australian Newsprint Mills, which has a Joint Venture scheme with farmers for pine plantations, has observed that farmers often approach him seeking diversification when there is a drop in the wool price [pers. comm.].

There are other products produced from trees apart from wood. These will be discussed in the following sections.

5.1. *Fodder trees and shrubs*

Fodder trees and shrubs can increase farm income from livestock by providing dry season feed. Compared to other purposes, only about 2% of farmers who practise agroforestry do so for fodder. This increases to 4% in the drier zones [Prinsley, 1991a]. Lefroy's review of fodder trees and shrubs (this issue) outlines the role that fodder trees can play in diversifying feed for livestock in the various agroecological zones of Australia and assesses their advantages and disadvantages. He also outlines the extent to which fodder trees occur and are used in Australia. Much of the following is extracted and modified from Lefroy's review.

5.1.1. *Nutritional advantages*

Fodder trees and shrubs have long been considered important for the nutrition of grazing animals in Australia, particularly in those areas with a pronounced dry season [Corbett, 1951; Imperial Agricultural Bureaux, 1947]. They provide a supplement of green feed when grasses and other herbaceous material are dry and they may provide the only source of protein and energy during drought.

5.1.2. *Nutritional disadvantages*

At the same time trees and shrubs have several disadvantages as sources of feed. They are often inaccessible to grazing animals. They are slow to establish, requiring isolation from stock during this time. Their foliage generally has higher fibre and lignin content than grasses [Wilson, 1969], and often has higher levels of tannins and other astringent compounds than shorter lived herbaceous plants. Although sometimes higher in protein, they often have lower energy value than herbaceous plants due to their lower digestibility. As a consequence of these factors they are rarely the first choice of grazing animals and seldom make up a significant proportion of the diet when grass and other herbaceous feed is available [Graetz and Wilson, 1979; Leigh and Mulham, 1986; Squires and Seibert, 1983].

5.1.3. *Environmental benefits*

Like all trees and shrubs, browse species can make a valuable contribution to landscape stability by decreasing the risk of wind and water erosion. They also contribute to the cycling of nutrients, especially nitrogen in the case of legumes, the cycling of water and to most of the other environmental benefits listed in Section 2 of this paper.

Browse plants have some role in grazing industries in all of Australia's major climatic zones (Fig. 3). Indigenous species are most important in the arid and semi-arid zones while cultivated species are being used in the monsoonal, sub tropical and temperate zones.

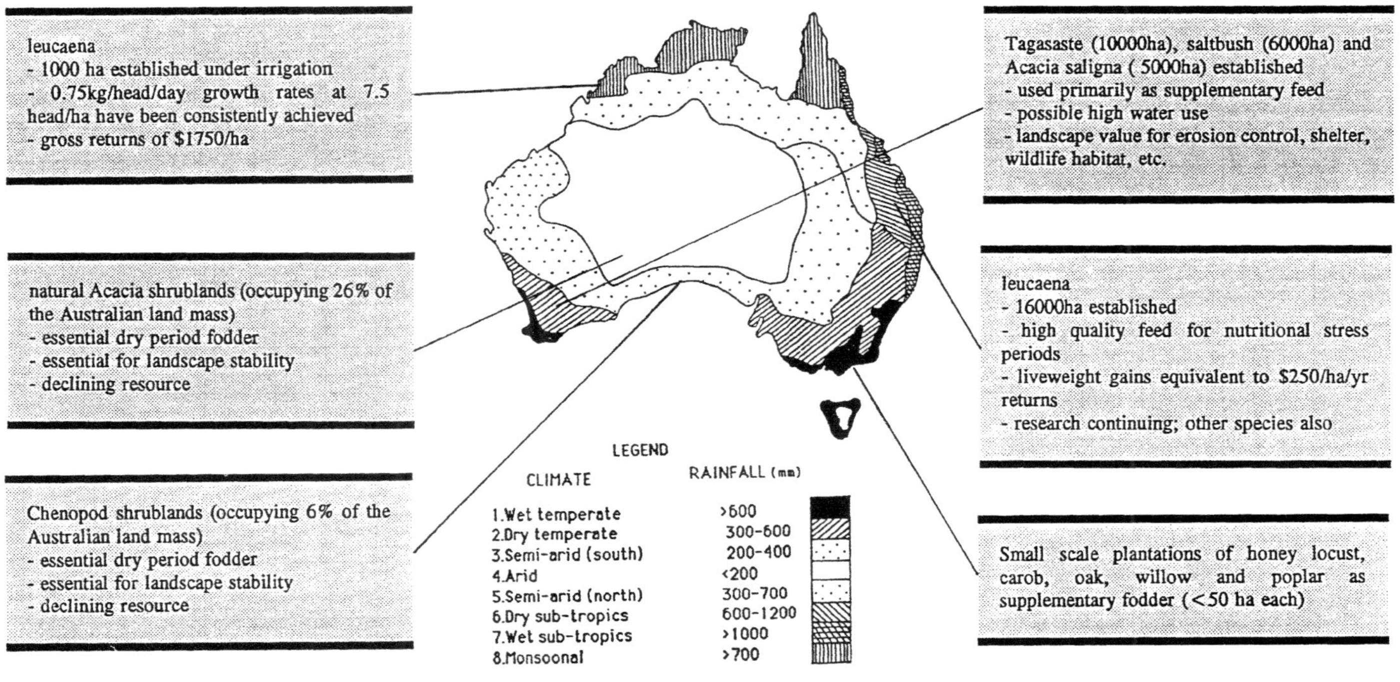

Fig. 3. Fodder trees and shrubs for each agroecological zone.

To assess the contribution of browse species to grazing systems, Lefroy considers their forage value to the animals, their economic value to the landholder and their ecological value in the landscape.

5.1.4. *Assessing the value of browse plants*

Increasing interest in the integration of trees and shrubs into agriculture is due in part to greater awareness of their protective role. This is particularly so in areas where agriculture has disrupted processes important in landscape stability such as the cycling of water and nutrients. In the case of browse plants, their productive and economic value needs to be assessed in comparison to the economic value of other options before it can be considered as an incentive to their wider use in a protective role. This productive role has two aspects; nutritional and economic. While a particular browse plant may be capable of maintaining or even fattening animals when handfed in an experiment, the cost of establishment and management in comparison to alternative feeds may not justify its wider use.

The place of cultivated browse plants is often limited by two factors. The marked seasonal nature of their feed advantage and the higher costs involved in growing and managing trees and shrubs compared to other forage plants. Together these limit the period of the year in which they have an economic advantage over other feed sources. In many cases this will restrict the area that can be profitably established to browse plants to that area sufficient to supply feed during a two to three month period each year, or about 10 to 20% of farm size. To plant a larger area requires a livestock trading enterprise that can take full advantage of an out-of-season feed supply. In summary, the value of trees and shrubs as fodder needs to be compared with other commercial uses when examining incentives for planting trees on farms.

Another limitation of the present use of cultivated browse species is the dependence on stands of single species. The use of mixed species of trees and shrubs and the selection of complementary inter-row grass and herbaceous species may lessen the impact of insects and disease.

5.1.5. *Research*

Management techniques must be developed to maximise production and persistence of cultivated browse plants. These may involve short periods of intense grazing followed by long periods of recovery. Strategies similar to this may be necessary in the arid and semi-arid rangelands to retain palatable species that are now in decline. There are currently 24 research projects across Australia investigating the use and management of fodder trees. This represents 10.7% of all agroforestry research in Australia [Prinsley, 1991a], and seems to reflect the relative importance of fodder trees in agricultural systems.

5.2. *Minor forest products*

The potential role for incorporating the production of minor forest products into normal farm management is an area which has received little attention to date. Very few indigenous plant species are commercially exploited in Australia [Cribb & Cribb, 1981], apart from those for timber and fodder uses. Equally neglected has been the production of minor forest products from non-indigenous species. In Boutland et al.'s review (this issue), they examine the position of products from trees, apart from timber, fodder and conservation plantings, which are potential sources of revenue.

Minor forest products which are significant in Australia include: essential oils, honey and pollen production, wildflowers and tree foliage, broombush, nut and other tree crops, sandalwood, cork, craftwood, seed, christmas trees, drugs, tannins, gums and resins, cane and charcoal. Many of these existing industries rely on exploiting natural resources, but there is increasing attention being paid to establishing plantations to produce these products. Some minor forest products may be produced Australia-wide (e.g. honey production), but in many cases products are restricted to the vegetation communities of particular climatic zones (eg. Queensland rainforests are the richest source of medicinal plants in Australia, Maiden, 1889).

Many potential industries have not succeeded due to their poor commercial prospects or to the lack of financial and/or research backing. There appear to be great disparities in the distribution and type of research afforded to these industries, with some having little or no research backing at all (e.g. craftwood) while others benefit from numerous research programmes (e.g. the Eucalypt oil industry). Much of the research that is being carried out is focussed on technical aspects of the industry or the resource, while little has been done on economics or marketing. Currently most minor forest product industries are not commercially viable as single enterprises primarily due to high production costs and overseas competition.

There appears to be some demand for minor forest products both overseas and in Australia. For minor forest products industries to become viable, enterprises need to be integrated with other commercial activities e.g. forestry, agriculture or other minor forest products. Likewise, opportunities for enhancing farm income solely from minor forest products are minimal. However, additional income could be made by incorporating appropriate species into multipurpose plantings.

In conclusion, Fig. 4 summarises Boutland et al.' paper, showing which products have potential for growing on farms in various parts of Australia. Their paper provides information concerning whether the industries shown in Fig. 4 are currently viable, potentially viable, or possibly viable.

5.3. *Conclusion*

Trees can produce many products which have the potential to benefit

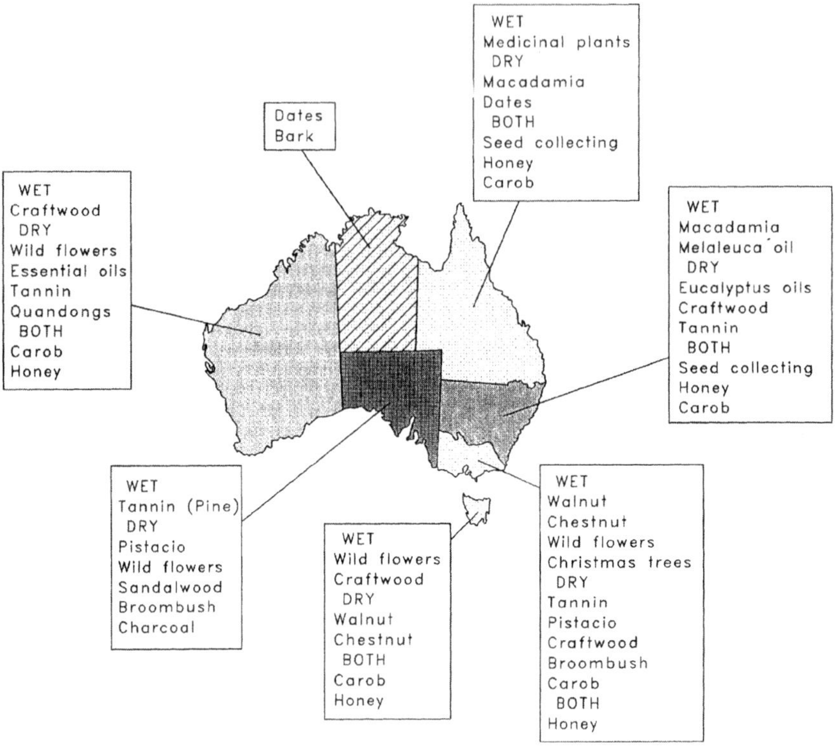

Fig. 4. Minor forest products which can be grown by famers in each State and Territory (information collated from Boutland et al., this issue) (wet is >600 mm rainfall/yr, dry is <600 mm rainfall/yr).

Australian agriculture. Most benefit will be obtained if trees are used for more than one purpose.

6. Discussion and conclusions

The successful adoption of agroforestry in Australia depends upon technical, economic, political and social factors. Most technical information is obtained through research via extension. The number of research projects being carried out in each State in Australia is known from a study carried out by Prinsley [1991a]. If this is compared with the number of farmers estimated to be practising agroforestry in each state then it is possible to see a correlation between ongoing research activities and agroforestry practices (see Fig. 5) for most states. Exceptions may be because some states have only recently

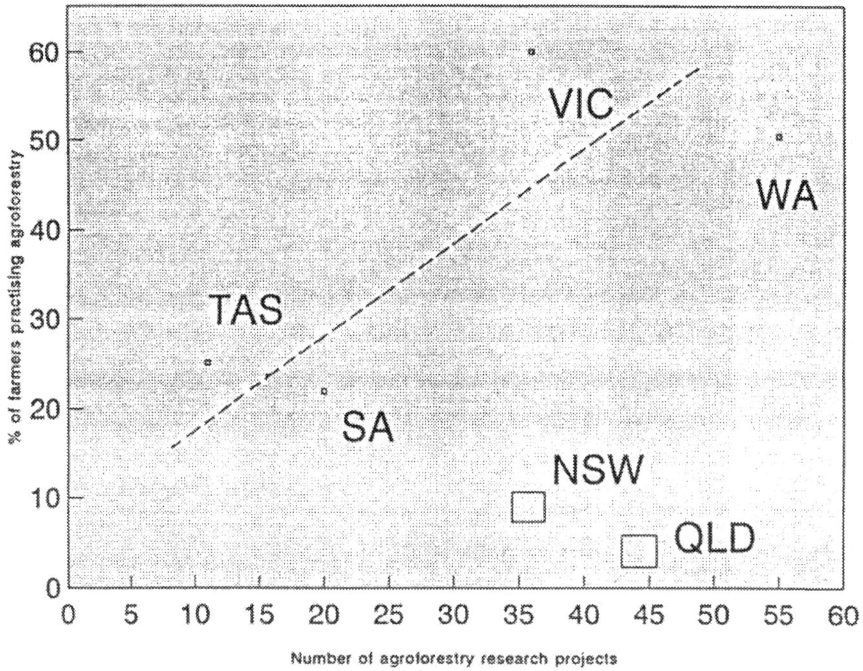

Fig. 5. Number of research projects and farmers practising agroforestry, by state.

started their research programs or because there are, as yet, limited specific extension or assistance schemes for agroforestry.

It is also possible to look at the numbers of agroforestry assistance schemes in each state as an indicator of the economic and political incentives for farmers to practise agroforestry. The numbers of these schemes [Prinsley, 1991a] appear to correlate with the percentage of farmers practising agroforestry in each state (see Fig. 6). These schemes are evaluated and described in Prinsley [1991b].

As pointed out by a number of review papers, (in particular, Schofield's) research regarded as 'successful' by scientists may not be implemented by farmers. Extension and other assistance agencies wishing to offer advice on technical or investment matters, are disappointed when these are not adopted by farmers. One reason why this happens is because the objectives and methods of scientists, extension and other agencies do not always match those of land users [Abel and Prinsley, 1991]. Farmers should be consulted by both research and assistance bodies in the formulation of research and extension strategies for their area. Successful consultation will help to ensure that research targets the needs of farmers and therefore that resources are efficiently used.

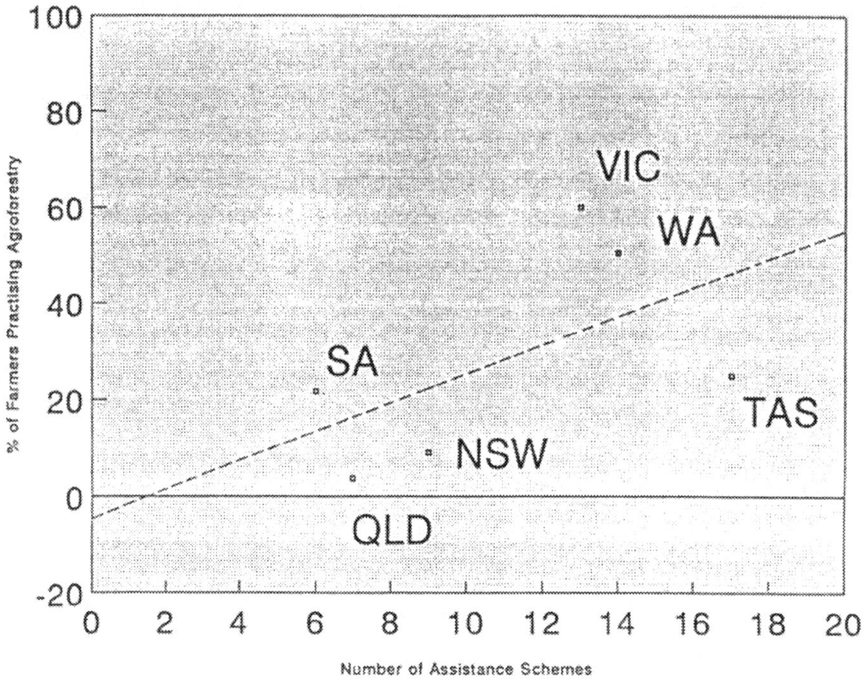

Fig. 6. Assistance schemes and the number of farmers practising agroforestry, by state.

Surveys of research priorities as perceived by farmers [Prinsley, 1991a] are useful, in conjunction with other information, for setting strategic directions for research on a state or national basis. However, when research is to be focussed upon a particular farming system or region within the broad outlines provided by such surveys, then in order to ensure that their research is appropriate, scientists should work in conjunction with farmers or community groups such as Landcare or Farm-Tree groups.

All new industries and land uses experience uncertainties at an early stage of their development. However this stage also provides the opportunity to develop the industry in a coordinated, and visionary way. The review papers in this journal are the first step towards such development in Australia. They should provide interesting and useful information from Australian research to scientists in other countries.

Acknowledgements

I would like to thank Andrew McNee and Mark Stafford-Smith for their assistance in drawing the maps and figures; Ann Hamblin, Mark Howden and Richard Moore for their comments on the paper and Jamie Allnutt for his assistance. I am grateful to all review authors from whom I received information.

References

Abel N and Prinsley RT (1991) Rapid appraisal for agroforestry research and extension. The Shuruguri Experience. Forest Ecology and Management, 45: 337—349

Australian Bureau of Agricultural and Resource Economics (1990) Commodity statistical bulletin. Canberra, Australian Government Printing Service

Bates CG (1911) Windbreaks: their influence and value. US Dept of Agric, Forest Services, Bulletin 86. Washington, Govt printing office

Bird PR (1988) Financial gains of trees on farms through shelter. In : Proceedings International Forestry Conference for the Australian Bicentenary 1988. Albury NSW, Australian Forest Development Institute, April 25—May 1, 1988, 11 pp

Bird JR and Cayley JWD (1991) Bad weather, shelter and stock losses. Agric Sci 4(4): 18—19

Bird PR, Lynch JJ and Obst JM (1984) Effect of shelter on plant and animal production. Proc Aust Soc Anim Prod 15: 270—273.

Bird PR, Bicknell D, Bulman P, Burke SJ, Leys J and Parker J (1992) The role of shelter in Australia for protecting soils, plants and livestock. Agroforestry Systems 20: 59—86

Boutland A, Byron N and Prinsley R (1990) 1991 Directory of assistamce schemes for trees on farms and rural vegetation. Greening Australia

Boutland A, Robinson M, Field J, Scheltema M, Hawkins B, Bulman P, Scarvelis J, Mason B, Wallace M, Castley M, Ryan P, Lansdown C and Reid R (1992) National Summary Paper — Minor Forest Products. Agroforestry Systems 20: 25—58

Bulman P (1991) Shade shelter and erosion control — a simplistic model for to compare the relative benefits of different farm tree designs for South Australia. The Role of Trees in Sustainable Agriculture conference, Sept 30—Oct 3, Albury, NSW

Caborn JM (1957) Shelterbelts and microclimate. Forestry Commission bull No 29, Edinburgh, UK

Cadman S, Prinsley RT and Moxham RW (1991) Environmental costs and benefits of establishing plantations on cleared agricultural land. Report to the National Plantations Advisory Committee. Bureau of Rural Resources, Canberra

Commonwealth of Australia (1979) A basis for soil conservation policy in Australia. Report no 1. Canberra, Australian Government Printing Service, 193 pp

Corbet HA (1951) Fodder trees: suggestions for their wider use. Perth, Imperial Printing Company

Cribb AB and Cribb JW (1981) Useful wild plants in Australia. Sydney, Collins

Eckersley R (1989) Regreening Australia: the environmental, economic and social benefit of reforestation. Occasional paper No 3. Melbourne, CSIRO

Ecos (1980) Soil Erosion: Can We Dam the Flood? Vol 25. Melbourne, CSIRO

Etherington D (1991) The economic analysis of trees in sustainable agriculture. Workshop paper for The Role of Trees in Sustainable Agriculture Conference, Sept 30—Oct 3, 1991, Albury, NSW

Ferguson IS (1988) Agroforestry as an economic venture in Australia. The International Forestry Conference, for the Australian Bicentenary, Vol 2. Albury, NSW, Australian Forest Development Institute

Geiger R (1965) The Climate Near the Ground. Massachusetts, Harvard Univ Press, 507 pp

Graetz RD and Wilson AD (1979) An assessment of herbivore diets in the chenopod shrublands. In: Graetz RD and Howes KMW, eds, Studies of the Australian Arid Zone. IV. Chenopod Shrublands, pp 144—159. Melbourne, CSIRO

Hall MJ (1990) Barriers and opportunities for growing plantations privately. In: Dargavel J and Semple N, eds, Prospects for Australain Forest Plantations, pp 389—96. Canberra, CRES ANU

Hansard AW and Dean MJ (1991) Taxation and industrial forestry investment: some issues and policy inplications, ABARE Discussion Paper 91.2. Canberra, AGPS

Imperial Agricultural Bureaux (1947) The use and misuse of shrubs and trees as fodder. Joint Publ. 10, Imperial Agr Bureaux, Aberystwyth

Ingram J (1990) The Role of Trees in Maintaining and Improving Soil Productivity — a Review of the Literature, Commonwealth Science Council, London

Institute of Forestry of Australia (1989) Trees: their key role in rural land management. Submission to the House of Representatives Committee of Inquiry into Land Degradation in Australia

Lefroy EC, Dann P, Wildin J, Wesley Smith R and McGowan T (1992) Trees and shrubs as a source of fodder in Australia. Agroforestry Systems 20: 116—138

Leigh JH and Mulham WE (1986) Selection of diet by sheep grazing semi arid pastures on the riverine plain. Aust J Exp Agric Anim Sci 6: 468—474

Loane B (1991) Economic Evaluation of Farm trees — Methodology and Data for Farmtree Model. Department of Agriculture & Department of Conservation and Environment

Maiden JH (1975) The Useful Native Plants of Australia. Compendium, Melbourne. Reproduction of 1889 edition

Marshall CJ (1990) Control of erosion. In: Cremer KW, ed, Trees for Rural Australia. Melbourne and Sydney, CSIRO, Inkata press

McComb JA, Bennett IJ, Bell DT and van der Moezel PG (1991) Adding value to Australia's genetic resources in tree species. Aust J Biotechnol (in press)

Moore R (1991) Producing timber from trees — options for farmers. The Role of Trees in Sustainable Agriculture Conference, Sept 30—Oct 3, Albury, NSW

Moore R (1992) Producing timber from trees — options for farmers. Agroforestry Systems 20: 165—184

National Farmers Federation (1991) Australian Agriculture. The Complete Reference on Rural Industry. Vol 2, 1989—1990. Canberra, NFF

Prinsley RT (1991a) Australian agroforestry — setting the scene for future research. Report to the Rural Industries Research and Development Corporation, Canberra, Australia

Prinsley RT (1991b) A review of private forestry assistance schemes on Australian farms. Report to the national plantations advisory committee. Bureau of Rural Resources, Canberra

Queensland Department Primary Industry (1990). 'Grassman'

Resource Assessment Commission (1991) Forest and timber inquiry draft, Report Vol 1, Canberra, Australia

Reid R and Bird R (1990) Shade and shelter. In: Kremer KW, ed, Trees for Rural Australia. Melbourne and Sydney, CSIRO, Inkata press)

Rural Research (1989) Wind erosion — dust, dunes,and tell-tale sediments, No 144, pp 13—23, CSIRO, Canberra

Scanlan JC (1992) A model of woody herbacious biomass relationships in eucalypt and mesquite communities. J Range Man 45 (Jan)

Scanlan J, Prinsley R, Piggott P, Prescott A, Bishop G, Wakefield S, Van der Sommen F, Duncan F, Stadler T, McLellan R and Ferago (1992) Management of native vegetation on farms in Australia. Agroforestry Systems 20: 139—164

Schofield NJ, Bari MA, Bell DT, Boddington WJ, George R and Pettit N (1992) Tree planting for dryland salinity control in Australia. Agroforestry Systems 20: 1—23

Squires VR and Siebert BD (1983) Botanical and chemical components of the diet of liveweight change in cattle on semi-desert rangeland in central Australia. Aust Rangel J (5)1: 28—34

Sturrock JW (1969) Aerodynamic studies of shelterbelts in New Zealand. 1. Low to medium height shelterbelts in mid Canterbury. NZ J Sci 12: 754—76

Sturrock J W (1972) Aerodynamic studies of shelterbelts in New Zealand. 2. Medium height to tall shelterbelts in mid Canterbury. NZ J Sci (15): 113—40

Walker J, Williams B, Hatton T, O'Loughlin E and Jupp D (1991) Use of trees for dryland salinity control. The Role of Trees in Sustainable Agriculture Conference, Sept 30—Oct 3, Albury, NSW

Wilson AD (1969) A Review of browse in the nutrition of grazing animals. J Range Man 22: 2328

Young A (1989) Agroforestry for Soil Conservation, Nairobi, ICRAF, CAB International, U.K. 121 pp

Agroforestry Systems **20**: 117–139, 1992.

Trees and shrubs as sources of fodder in Australia

E.C. LEFROY[1], P. R. DANN[2], J. H. WILDIN[3], R.N. WESLEY-SMITH[4] and A. A. McGOWAN[5]

[1] *Western Australian Department of Agriculture, Perth;* [2] *CSIRO, Canberra;* [3] *Department of Primary Industries, Rockhampton, Queensland;* [4] *Department of Primary Industries and Fisheries, Berrimah, Northern Territory;* [5] *Victorian Department of Agriculture, Ellinbank*

Key words: browse, *Leucaena, Chamaecytisus, Atriplex, acacia*

Abstract. Experience with browse plants in Australia is briefly reviewed in terms of their forage value to animals, their economic value to the landholder and their ecological contribution to landscape stability. Of the cultivated species only two have achieved any degree of commercial acceptance (*Leucaena leucocephala* and *Chamaecytisus palmensis*). Both of these are of sufficiently high forage value to be used as the sole source of feed during seasonal periods of nutritional shortage. Both are also leguminous shrubs that establish readily from seed. It is suggested that a limitation in their present use is the reliance on stands of single species which leaves these grazing systems vulnerable to disease and insects. Grazing systems so far developed for high production and persistence of cultivated species involve short periods of intense grazing followed by long periods of recovery. Similar management may be necessary in the arid and semi-arid rangelands where palatable browse species are in decline.

1. Introduction

Trees and shrubs, often called browse or topfeed, have long been considered important for the nutrition of grazing animals in Australia, particularly in those areas with a pronounced dry season [3, 14]. They provide a supplement of green feed when grasses and other herbaceous material is dry and they provide the only source of protein and energy during drought when all other feed is absent.

At the same time trees and shrubs have several disadvantages as sources of feed. They are often inaccessible to grazing animals. They are slow to establish requiring isolation from stock. Their foliage generally has higher fibre and lignin content than grasses [52], and often has higher levels of tannins and other astringent compounds than shorter lived herbaceous plants. Although sometimes higher in protein, they often have lower energy value than herbaceous plants due to their lower digestibility.

As a consequence of these factors they are rarely the first choice of grazing animals and seldom make up a significant proportion of the diet when grass and other herbaceous feed is available [12, 19, 42].

To assess the contribution of browse species to grazing systems it is necessary to consider in turn their forage value to the animals, their economic value to the landholder and their ecological value in the landscape.

2. Assessing the value of browse plants

Increasing interest in the integration of trees and shrubs into broadacre agriculture is due in part to greater awareness of their protective role. This is particularly so in areas where agriculture has disrupted processes important in landscape stability such as the cycling of water and nutrients.

In the case of browse plants, their productive value needs to be carefully assessed before it can be considered as an incentive to their wider use in a protective role.

This productive role has two aspects; nutritional and economic. While a particular browse plant may be capable of maintaining or even fattening animals when handfed in an experiment, the cost of establishment and management in comparison to alternative feeds may not justify its wider use.

The place of cultivated browse plants is often limited by two factors. The marked seasonal nature of their feed advantage and the higher costs involved in growing and managing trees and shrubs compared to other forage plants. Together these limit the period of the year in which they have an economic advantage over other feed sources. In many cases this will restrict the area that can be profitably established to browse plants to that area sufficient to supply feed during a two to three month period each year, or about 10 to 20% of farm size. To plant a larger area requires a livestock trading enterprise that can take full advantage of an out-of-season feed supply.

Whichever case, the value of trees and shrubs as fodder needs to be kept in perspective with other commercial end uses when examining incentives for planting trees on farms.

2.1. *Forage value*

The forage value of any feed depends on the combination of its palatability, nutritive value and digestibility. These need to be sufficiently high for an animal to take in its daily requirement of energy, protein and minerals. That minimum requirement varies with the type of animal, the desired result (maintenance, weight gain, wool growth, milk production) and climatic conditions. For a 50 kg wether for example, the daily maintenance requirement is eight megajoules of metabolizable energy, 95 g of crude protein, 5 g of calcium, 1 g of magnesium and 2 g of sodium [39]. To acquire this amount of energy means taking in 500 g of digestible organic matter, or 1.1 kg dry weight of feed that is 55% digestible.

The intake of sufficient energy and nutrients by an animal cannot be predicted from separate analysis of a plant's nutrient content, digestibility or palatability. Despite this, the wider use of trees and shrubs as forage is often advocated on the basis of such analyses [3, 4, 14, 45]. While these can serve as some guide to the value of species, they must be regarded with caution for several reasons.

Firstly, chemical analysis commonly overestimates digestibility, particularly that of protein, as it does not take into account the fact that protein is often bound to lignin and tannins which can prevent its breakdown in animals [6, 27]. Secondly, digestibility can be a poor indicator of forage value as shown in a study of seven common Australian browse plants where there was no relationship between true digestibility and the amount voluntarily eaten by sheep [53]. Thirdly, palatability can vary seasonally and between animals and cannot therefore be assessed on the basis of the occasional consumption of browse [6, 52].

Even when pen feeding experiments are carried out, the results can be misleading with browse plants. In such experiments, the material is collected by hand so the results are insensitive to the accessibility of the feed. Accessibility is not only a function of height, but also of the spatial distribution or density of the edible material, and therefore the amount of foraging required by the animal to take in sufficient feed.

For these reasons, an accurate assessment of the forage value of browse species can only be made from the response of grazing animals. This paper concentrates on species for which such information is available. A summary of the factors important in assessing forage value is given in Table 1.

Table 1. Assessing the value of browse plants.

I. Forage value to the animal	II. Economic value to the landholder	III. Landscape value to the farm or catchment
Is the plant material:		Does the plant match site requirements for:
• accessible		
• acceptable	• establishment cost	• reduced wind erosion
• digestible (> 55% DMD)*	• management cost	• reduced water erosion
• sufficiently high in	• time to first use	• water use
— energy	• substitution value	• increased nutrient cycling
— protein	• persistence	• improved soil structure
— minerals	• complementary effects	• increased genetic diversity
• non-toxic	• competitive effects	• decreased energy inputs
Measure: response in grazing animals	Measure: return on investment	Measure: landscape stability

* DMD is dry matter digestibility.

2.2. *Economic value*

Assuming that a browse plant can meet the required forage value, its economic value to the landholder will depend on some additional characteristics of the plant; the cost of establishing and managing the trees and shrubs, the time taken for them to reach a productive age, the persistence of the plant under regular use (in terms of both its ability to regrow each year and its longevity), the ability of the plant to produce feed when it is most needed.

These attributes can be expressed in terms of its substitution value, meaning the cost of producing browse feed compared to the cost of alternative feed sources necessary to achieve the same animal response. By taking into account the expected life of the plants, this value can also be expressed as the rate of return on investment.

Assessment of economic value also needs to take into account any complementary or competitive effects of trees and shrubs. An example of a complementary effect is deferred grazing; by having browse feed available during a time of seasonal shortage, other areas of land can be relieved of grazing pressure and given an opportunity to consolidate. Another complementary effect is the increased growth of herbaceous species growing amongst browse plants due to shelter or shade effects [15, 44].

Competitive effects include the lost production from herbaceous plants on the land occupied by the browse species. A summary of the factors important in economic assessment is shown in Table 1.

2.3. *Landscape values*

Like all trees and shrubs, browse species can make a valuable contribution to landscape stability by decreasing the risk of wind and water erosion. They also contribute to the cycling of nutrients, especially nitrogen in the case of legumes, and the cycling of water.

Their deeper and more permanent root systems also help to overcome problems of declining surface structure and compaction commonly associated with crop and pasture land. The introduction of browse plants can also increase the genetic diversity in the landscape and therefore help to increase the resilience of the vegetative cover in general to pests, diseases, fire and climatic extremes.

The use of perennial plants can also increase the energy efficiency of agriculture by reducing the requirement for cultivation, and in the case of legumes, reduced input of fertilizer.

Browse plants are becoming important in parts of Australia because they bring these attributes to land that is unproductive under conventional agriculture or is degraded due to erosion or salinity. A summary of these landscape values is given in Table 1.

2.4. *Distribution of species by climatic zones*

Figure 1 shows Australia divided into the major climatic zones. Browse plants have some role in grazing industries in all of these zones. Indigenous species are most important in the semi-arid and arid zones while cultivated species are being used in the monsoonal, subtropical and temperate zones.

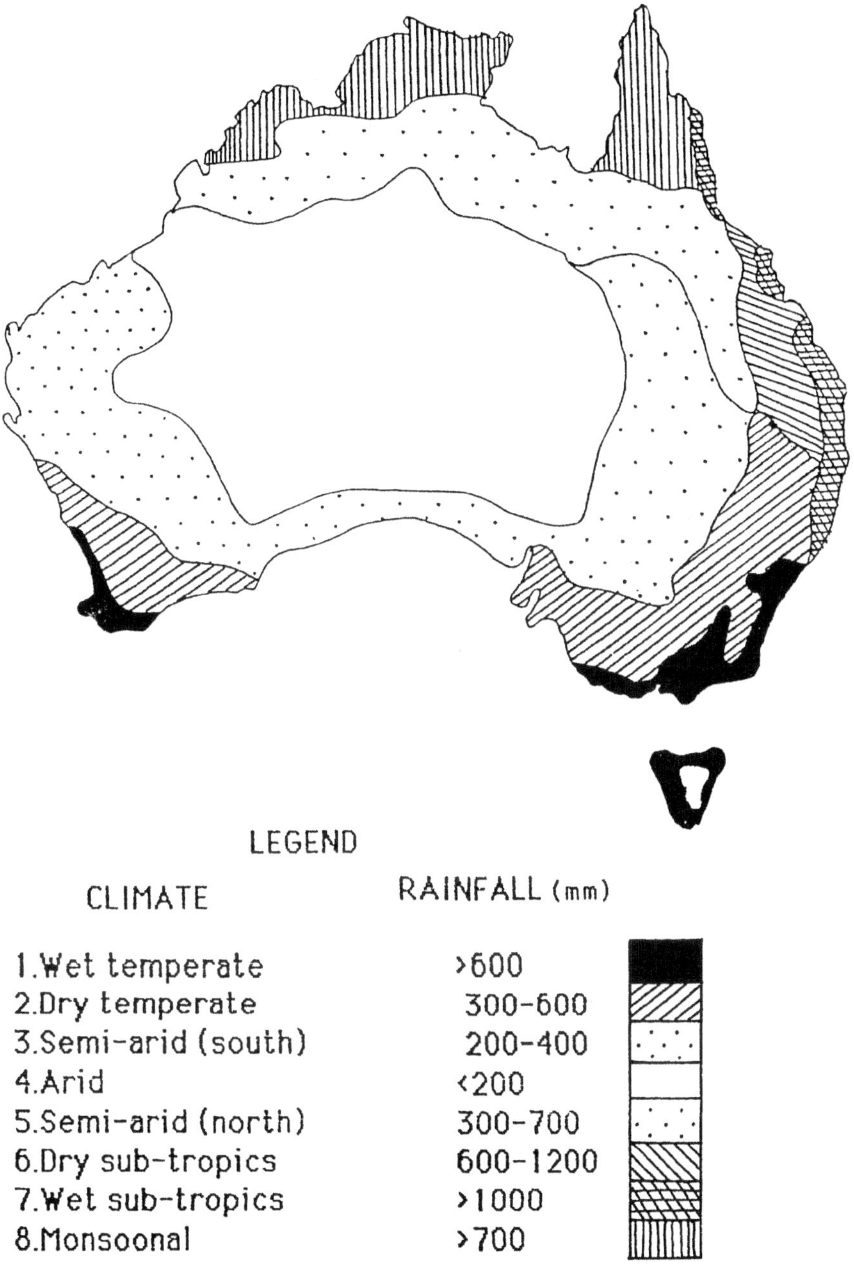

Fig. 1. The major climatic zones of Australia.

3. Arid and semi-arid rangelands

3.1. *The role of browse plants in the rangelands*

Over 70% of Australia falls within the arid and semi-arid climatic zones. The dominant land use is extensive grazing by sheep and cattle at stocking rates that vary from the equivalent of 1 sheep to 4 ha to 1 sheep to 40 ha. This wide range in carrying capacity reflects variation in vegetation type and it is within three main vegetation types that edible trees and shrubs are found. These are the Acacia shrublands occurring mainly in the south and centre, the low Chenopod shrublands of the south and the semi-arid woodland to the east and north. The remaining important vegetation type of this region is grassland.

The first two of these vegetation types (Acacia and Chenopod shrubland) occupy 26% and 5% of the Australian land mass, respectively [43]. Table 2 lists some of the browse plants of this region.

Despite historical reference to their general value to the grazing industry [3, 9, 14] it is apparent from more recent studies that their contribution to animal production in most seasons is very small as both sheep and cattle heavily favour annual and perennial grasses and forbs due to their consistently higher digestibility. In particular, little or no relationship has been shown between a high density of browse plants and high animal productivity in the riverine plain of eastern Australia [11, 12, 19]. Similarly a comparison of rangeland with low and high density of palatable shrubs in Western Australia showed no difference in wool production over five years except at very high stocking rates in a drought year. At average to low stocking rates (6 to 12 ha per sheep) there was no difference in wool production as the area with less shrubs produced three to four times the amount of herbaceous feed [47].

In central Australia, studies of the botanical composition of the diet selected by cattle have shown that browse usually comprises less than 10% of the diet but may rise to 20% during very dry times [2, 42] (see Fig. 2).

In short it is fairly well established that the major value of browse species to the grazing industries in the arid and semi-arid zones is in ensuring continuity of animal production over very dry periods. In addition, studies of land degradation in the rangelands have concluded that shrubs, through their roles in nutrient cycling and erosion prevention, are important in preserving landscape stability [24, 50]. However, no detailed studies to date have been of long enough duration to demonstrate an economic benefit in this regard.

For these reasons it is of concern that the density of palatable browse species is decreasing in both the Chenopod shrublands and the Acacia shrublands, that grazing pressure plus the exclusion of fire is causing grasses to be displaced by less desirable woody plants in semi-arid woodlands and that levels of erosion are increasing in all three [48, 54].

The apparent lack of economic incentive to reverse these trends means

Table 2. Some browse plants of the arid and semi arid regions.

	Common name	Botanical name	Digestibility (%)	Crude protein (%)
Trees	Bowgada (pods and seeds)	*Acacia linophylla*	–	13–20
	Currara	*Acacia tetragonaphylla*	–	6–16
	Kurrajong	*Brachychiton populneum*	42–61	–
	Mulga	*Acacia aneura*	35–64	6–20
	Supplejack	*Ventilago viminalis*	–	–
	Whitewood	*Atalaya hemiglauca*	35	–
	Wilga	*Geijera parviflora*	48–62	12
	Witchetty bush	*Acacia kempeana*	58	–
Shrubs	Bladder saltbush	*Atriplex vesicaria*	–	12
	Bluebush	*Maireana polypterygia*	60–74	15–19
	Cottonbush	*Ptilotus obovatus*	41–58	8–14
	Old Man Saltbush	*Atriplex nummularia*	52–67	14
	Rhagodia	*Rhagodia eremea*	67	6–12
	River saltbush	*Atriplex amnicola*	65–70	9–10

Sources: [7, 10, 14, 28, 46].

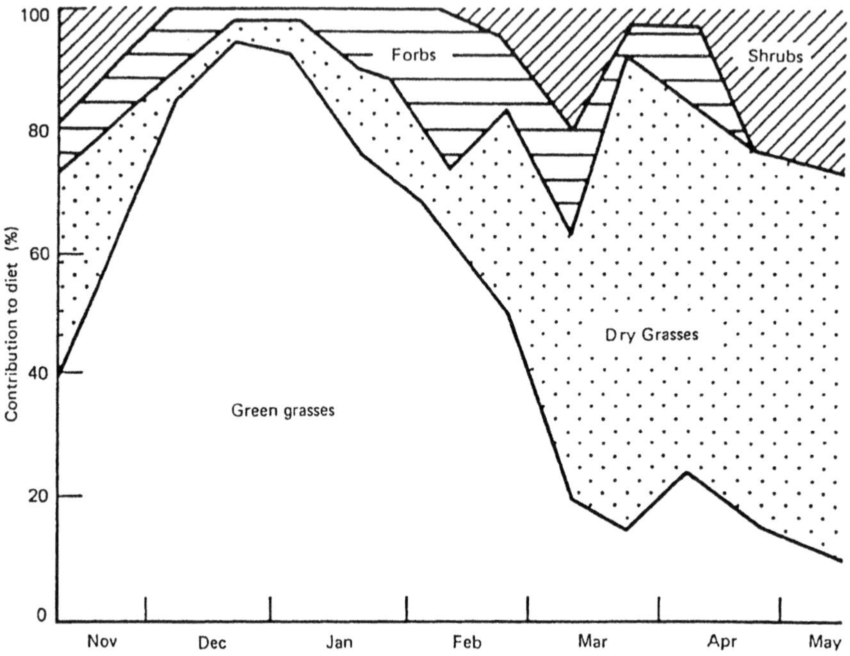

Fig. 2. Proportions of green grass, dry grass, forbs and shrubs in the diet of steers grazing between November 1977 and May 1978, Alice Springs (from Squires and Siebert, 1983 [42]).

that the preservation of range condition is seen to be largely dependent on the development of a conservation ethic among landholders. To this end, monitoring sites are being established throughout Western Australia and South Australia. These are intended to both detail changes in range condition, and through regular photography to have a graphic educational role [13].

3.2. *The future of browse plants in the rangelands*

The retention of desirable species in the rangelands, both in terms of their ecological functions and their value for animal production, will depend on an understanding of the interactions between several factors. Important among these are grazing intensity, grazing frequency, location of watering points, the use of fire, the populations of native and feral animals and local variations in climate and soil type.

While theoretical frameworks put forward by ecologists can be useful in developing management prescriptions [40], the long-term experience and observations of landholders are an invaluable source of information that has possibly been under-utilized. This is particularly so because their experience is based on observing the interactions mentioned above at the scale of operations most relevant to other managers; that of an entire pastoral lease and not one paddock or other restricted study area.

Two published examples of such work [29, 34] offer examples for future co-operative study into management strategies. In the first case, observations over 28 years have shown that a conservative stocking rate and a rotational grazing system designed to increase recruitment of chenopod shrubs have resulted in an increase in shrub density and patterns of wool production and lambing percentage that are relatively independent of rainfall. A survey of 12 nearby pastoral leases showed this one to have the lowest overall stocking rate (1 sheep to 21 ha compared to 1 to 13) and the highest income per hectare. This last point suggests that financial incentives as well as a growing land ethic may contribute to the development of sustainable grazing systems in some areas.

4. Cultivated species

The total area of trees and shrubs cultivated for fodder in Australia amounts to between 20,000 and 30,000 ha only. This is divided between the subtropical north-east coast and the dry temperate south-west. Small areas have also been planted in the temperate south-east and under irrigation in the north-western monsoonal zone.

Over three-quarters of the total area is accounted for by two species; leucaena (*Leucaena leucocephala*) in the north and tagasaste (*Chamaecytisus palmensis*) in the south.

Both are multi-branched leguminous shrubs that appear to be vigorous early colonizers of disturbed ground in their respective habitats. Smaller areas of various saltbush (*Atriplex* spp.), Acacias, poplars (*Populus* spp.), willows (*Salix* spp.) and carob (*Ceratonia siliqua*) have been planted in southern Australia (Table 3). Work is in progress at several institutions throughout Australia to select further species, primarily for the sub-tropics (Table 4).

4.1. *Subtropical north-east*

Leucaena (*Leucaena leucocephala*)

Some 16,000 ha of leucaena have been planted in the last 10 years in areas receiving between 500—1000 mm annual rainfall. The high forage quality of its foliage has been well established [16] as has its ability to produce high liveweight gains in cattle [35].

Its current role in dryland cattle production is to supply high quality feed during the traditional period of nutritional stress from April to October when other pasture is of low quality. In this way cattle are being fattened in areas

Table 3. Trees and shrubs cultivated for fodder in Australia.

Common name	Botanical name	Climatic zone	Estimated area (hectares)
Browse shrubs			
Leucaena	*Leucaena leucocephala*	Sub-tropics	16,000
		Monsoonal	1,000
Tagasaste	*Chamaecytisus palmensis*	Dry temperate	10,000
Saltbush	*Atriplex* spp.	Dry temperate	6,000
Wattle	*Acacia saligna*	Dry temperate	5,000
Trees for foliage			
Willow	*Salix* spp.		
Poplar	*Populus* spp.		
Paulownia	*Paulownia* spp.		
		wet temperate	< 50 each
Trees for pods or fruit			
Oak	*Quercus suber*		
Carob	*Ceratonia siliqua*		
Honey locust	*Gleditsia trichanthos*		

Table 4. Some species of fodder trees and shrubs currently under investigation in Australia.

Species	Institution*
Acacia angustissima	QDPI/CSIRO
Acacia aneura	QDPI (Chareville)
Acacia microbotrya	WADA
Acacia saligna	WADA
Albezia chinensis	QDPI/CSIRO; University; of Queensland
Albezia lebbeck	QDPI/CSIRO
Atriplex amnicola	WADA/DARA
Atriplex lentiformis	WADA/DARA
Atriplex undulata	WADA/DARA
Bitumenaria bitumenosa	WADA
Cajanus cajan	QDPI/CSIRO; University of Queensland
Callianadra calothyrus	QDPI/CSIRO; University of Queenland
Chamaecytisus palmensis	MRP: DARA: NSWAF
Cytissus mollis	WADA
Gliricidia sepium	QDPI/CSIRO; University of Queensland
Leucaena diversifolia	QDPI/CSIRO
Leucaena leucocephala	QDPI/CSIRO; University of Queensland; WADA
Leucaena pallida	QDPI/CSIRO
Medicago arborea	WADA
Sesbania sesban	University of Queensland; QDPI
Teline monspessulana	WADA

QDPI = Queensland Department of Primary Industries.
CSIRO = CSIRO Division of Tropical Crops and Pastures.
WADA = Western Australian Department of Agriculture.
MRP = Martindale Research Project, University of Western Australia.
DARA = Department of Agriculture and Rural Affairs, Victoria.
NSWAF = New South Wales Department of Agriculture and Fisheries.

previously only suitable for breeding. Liveweight gains of 200—300 kg/ha per year have meant gross returns of about $250/ha from leucaena at current prices. Establishment costs are in the order of $200—250/ha [51].

Under dryland conditions, establishment is by direct seeding in rows 5 to 10 m apart with first grazing at 12 months. Grasses such as buffel (*Cenchrus ciliaris*) and panic (*Panicum maximum*) are sometimes sown in the inter-row 12 months after establishment. Cutting is unnecessary as grazing alone will keep all regrowth within the reach of cattle. Under irrigation, denser stands sown in rows 2 m apart have produced liveweight gains of about 1000 kg/ha/year.

Problems were initially encountered with the side effects of the high mimosine content of leucaena. The breakdown of this amino acid to the toxic compound 3 hydroxy-4-(IH) pyridone (DHP) resulted in hair loss, weight loss and in some cases death which limited the role of leucaena to that of a supplement. A bacteria capable of degrading DHP has since been introduced

Fig. 3. Cattle grazing irrigated leucaena (*Leucaena leucocephala*). Photo: D. Pratchett.

to most districts and leucaena now constitutes the sole feed source in many cases.

The arrival of the leucaena psyllid (*Heteropsylla cubana*) in 1986 was a cause of concern however it now appears that the damage caused by this leaf-sucking insect is restricted to more humid coastal areas.

However, selecting for psyllid resistance is seen as a high priority for research. Other areas seen as requiring work are selection for early growth and hence better establishment, selection for frost resistance to extend its range and selection of other species to limit the risk of heavy dependence on one plant to fill this role.

4.2. *Monsoonal north-west*

About 1000 ha of leucaena is being grown under irrigation on heavy clay soils of the Ord Irrigation Area for fattening cattle [32]. Growth rates of 0.75 kg/head/day have consistently been achieved at stocking rates of 7.5 animals per hectare (Fig. 3). The current system of grazing involves rotating cattle at an average stocking rate of 7.5 per hectare through paddocks with rows of leucaena planted 3 m apart. Paddocks are grazed for one week and allowed three weeks to recover before next use.

Range reared cattle are presently being fattened from 150 kg to 300 or

Fig. 4. Young sheep grazing tagasaste (*Chamaecytisus palmensis*). Photo: C. M. Oldham.

400 kg depending on the market. The average gross return per hectare in 1990 was $1750 [33]. Land preparation and seeding costs range from $200 to $650/ha.

Establishment is seen as an area for further work with weed competition and low seedling vigour under these conditions the major problem. High stocking rates on irrigated clays means that compaction may cause damage to soil structure and plant roots and be a major limitation of this system. A high priority is to find an inter-row grass that will establish easily from seed.

4.3. *Dry temperate zone*

This winter rainfall zone takes in the cereal and sheep growing area of southern Australia receiving between 250 and 600 mm of rainfall annually. Cereal crops are grown in rotation with annual pasture or grain legume crops. The temperate zone supports the majority of Australia's 200 million sheep. Many of these are fed supplements of grain for two to three months over autumn and early winter each year. This supplementation fills the gap between the last of the crop residues plus dry annual pasture and the first growth of the next season's pasture.

Three cultivated browse plants are presently being used on a small scale as substitutes for this supplementary feeding; tagasaste (*Chamaecytisus pal-*

mensis), saltbush (various species of *Atriplex*) and acacia (*Acacia saligna*). There is also interest in these fodder shrubs for their potential ability to increase water use and hence overcome problems of rising water tables and increasing secondary salinity. They also have a role to play in stabilizing areas affected by erosion, in sheltering stock from adverse weather conditions and in providing habitat for beneficial insects and wildlife in general. It is their value as forage however that will act as a primary incentive to their wider use.

Tagasaste (*Chamaecytisus palmensis*)
This shrub is native to the arid slopes of the island of La Palma in the Canary Islands and was first introduced to Australia in 1879. It is now naturalised in Southern Australia on freely drained soils in areas receiving more than 500 mm of annual rainfall. It is intolerant of frost, waterlogging or salinity [38]. Some 10,000 ha were planted on predominantly deep sandy soils on farms in Western Australia between 1985 and 1990 (Fig. 4). Smaller experimental areas have been planted in Victoria, New South Wales and South Australia.

Forage value of tagasaste. This shrub grows vigorously on deep infertile soils yielding 3 tonnes per hectare of edible dry matter each year on 450 mm of rainfall [7, 31], and up to 9 tonnes per hectare on 1100 mm rainfall [26]. The combination of the shrubs and the inter-row pasture brings production in line with the generally accepted relationship between effective rainfall and forage production in this climatic zone of around one tonne per hectare for every 100 mm of rain [49].

In summer, the edible fraction (leaf plus stem <5 mm) of tagasaste contains about 15% protein and is about 70% digestible. It will regrow rapidly after cutting or grazing in summer without rainfall. As it tends to grow into a spreading shrub up to 5 m high, it has been found necessary to mechanically prune trees grown under cultivation. Trees are generally grown in rows 5 to 10 m apart at densities of about 1000 to 2000 trees per hectare and cut once a year prior to grazing.

The current pattern of use involves grazing shrubs once a year over a 30—60 day period at high stocking rates (up to 50 sheep per hectare) followed by a 10 month rest period over winter and spring.

With 450 mm rainfall per year, this regime has produced a yield of about 3000 sheep grazing days per hectare per year. Tagasaste used in this way has maintained sheep in the same or better condition as dry annual pasture plus grain supplement over five years of experiments [31]. With higher rainfall, it is likely that the frequency of grazing or harvesting can be increased.

Growth rates of sheep and cattle on tagasaste are much less than estimates suggested by chemical analysis [7, 26] but it is capable of reducing the loss of condition normally experienced in autumn. From a starting weight of 32 kg, young sheep in one experiment lost 2 kg over a 20 week period on tagasaste compared to 5 kg over the same period on dry pasture and grain supplement [31].

With cattle, liveweight gains have been insignificant when they are forced to eat all the browse available in a paddock. When moved regularly onto fresh paddocks, they have shown liveweight gains of 500 to 600 g per day. This suggests that sequential use by cattle and sheep may produce liveweight gains in cattle and supply maintenance for sheep.

The ability of tagasaste to grow extra wool compared to dry pasture and supplements has been demonstrated in three consecutive years with young sheep. Clean fleece weights were increased by 25 to 30%; however, the quality and subsequently the value of the extra wool has been inconsistent [31].

Economic value of tagasaste. Tagasaste can profitably substitute for hand feeding of grain as maintenance feed for young sheep with returns on investment of 23% at a yield of 3400 sheep grazing days per hectare and 10% at 2170 sheep grazing days per hectare [25]. This analysis was based on the actual establishment, management and yield of a commercial plantation, including the cost of fencing and cutting, and assuming the plantation has a life of 15 years.

It has been shown that the efficiency of utilization of this feed source can be further improved by ration grazing. Sheep have been maintained in similar condition over autumn whether they had access to tagasaste for seven days a week, three days a week and only one day a week, with the three groups showing no significant difference in clean fleece weight or quality [31].

The profitability of using tagasaste to grow extra wool over summer and autumn on tagasaste is less conclusive. Although wool growth rates can be maintained at higher levels than that of flockmates on dry summer pasture, inconsistent effects on mean fibre diameter have meant returns have varied from a loss in one year to an increase in returns of 15% in other years.

Tagasaste is usually established using specialized tree seeders. These machines rip, scalp a 1 m width of topsoil to remove weeds and weed seed, drop about 10 seeds per metre of row and follow with a press wheel. The cost varies from $80 to $150 per hectare. Maintaining shrubs at grazing height requires annual pruning at $30/ha.

Landscape value of tagasaste. Tagasaste has shown it can fill an important gap in current land use patterns. There are several million hectares of infertile sandy soils in southern Australia. While 600,000 ha of the more fertile of these sandy soils are currently used for the production of grain lupins, much of this area has proved unsatisfactory for annual pasture or cereal growing and remains susceptible to wind erosion and in many cases contributes to recharge of ground water and subsequent valley floor salinity.

Although it has been shown that tagasaste can be more productive than alternative uses of these soils in a 450 mm rainfall area, the role it can play is limited by the fact that it only appears profitable to plant up to the point that the autumn feed gap is filled. On most farms this amounts to about 10% of farm size [31].

Further research may demonstrate it has a profitable role in increasing the year round carrying capacity of sheep or cattle.

Other perennial species are currently being screened for the temperate zone (see Table 4). In this way it is hoped to increase the range of browse plants available to reduce the risk of depending on one species and to select species that do not require pruning. Perennial grasses may also have a role in complementing the feed of browse plants. Preliminary investigations with perennial veldt grass (Erharta calycina var. "Mission") and Rhodes grass (*Chloris gayana* var. "Pioneer") suggest they can increase the carrying capacity in years with summer rain when sown between rows of tagasaste [C.M. Oldham pers. comm.].

Saltbush (*Atriplex* sp.)

Over one million hectares of land have become saline this century in southern Australia as a result of clearing for agriculture. Selection of salt tolerant plants from Australia and overseas for their ability to produce forage on saline land has resulted in about 6000 hectares of shrubs established on farms in the last decade.

Those indigenous to the salt lakes and drainage lines of dry temperate and semi-arid Australia include River Saltbush (*Atriplex amnicola*), Grey Saltbush (*A. cinerea*), Old Man Saltbush (*A. nummularia*) and small leaved bluebush (*Maireana brevifolia*). Two well adapted exotic species are Quail Brush from southern United States (*A. lentiformis*) and Wavy Leaved Saltbush from Chile (*A. undulata*) [21].

Two other indigenous species, Samphire (*Halosarcia* spp.) and Marsh Saltbush (*A. paludosa*), often spread naturally in saltland managed for grazing and contribute to the feed supply but are rarely planted [22].

Forage value of saltbush. Commercial experience has shown that saltbush and bluebush pastures can be used to maintain the condition of sheep over summer and autumn as a substitute to grain feeding [22].

There is, however, little experimental work comparing the nutritional value of saltbush with alternative feeds. One study has indicated there are disadvantages with a pure saltbush diet compared to a mixed diet of saltbush and hay. Sheep fed *Atriplex undulata, A. lentiformis, A. amnicola* and *A. cinerea* diets lost weight at about 200 g per day while sheep on an equal mixture of oaten hay and *A. undulata* gained 70 g per day over the three week trial [46].

The likely advantages of the mixed diet are a lowered salt intake than the pure saltbush diet and higher crude protein than the pure hay. The high salt content of the saltbush diets (150—180 g total ash per kg dry matter) may have meant that sheep reached their upper daily limit of salt before there had been sufficient intake for maintenance.

In addition, an average water intake of 8 litres per kilogram dry matter of saltbush may be responsible for lowering the efficiency of digestion [1]. The two species with highest sodium, potassium and total ash content (*A. amnicola* and *A. cinerea*) had significantly lower nitrogen digestibility in this experiment.

This suggests the role of saltbush is as a supplement of protein and other nutrients normally at low levels in dry grass and crop residues. In that case the establishment of companion grasses in saltland such as *Puccinellia ciliata* and *Agropyron elongatum* becomes equally important, as does the grazing of saltland in conjunction with crop residues and dry pasture.

Accessibility limits the forage value under grazing of two species, *A. nummularia* and *A. lentiformis*, both exceeding the 1.2 m browsing height of sheep. For this reason they are often planted as a small proportion within a mixture of shrubs.

The forage value of saltbush may also be limited by the presence of oxalates which are toxic to ruminants. Levels of about 6% in *A. amnicola* and *A. undulata* and 9—14% in *Maireana brevifolia* have been recorded [46].

Economic value of saltbush. Under commercial practice sheep grazing saltbush pastures select a mixed diet that includes grass and other herbage which would complement the shrub forage and offset the ill effects of a high saltbush diet [12, 19, 20].

Estimates to date of the profitability of saltbush under cultivation have been based on the assumption that saltbush is the sole source of nutrition and have assumed levels of dry matter intake (and subsequently metabolizable energy) that have not been born out in practice [37].

From commercial practice it is reported that stands of saltbush plus volunteer herbage can provide up to 2500 grazing days per hectare per year and maintain the condition of sheep [22]. In such cases the value of saltbush can be expressed in terms of the cost of alternative feed. Hand feeding of grain currently costs about $1 per head per month and would be necessary for two to three months each year. On that basis saltbush pastures yielding 2500 grazing days per hectare per year return between $28 and $42 per hectare annually.

Saltbush pastures are currently established from seed using niche seeding techniques [23]. Seeds are sown in mounded rows about 3 m apart with 2 to 3 m between plants in a row. The cost of establishment is between $200 and $280 per hectare.

As saltbush and bluebush both recruit from seed or by layering, the lifespan of individual plants is not important for the persistence of a pasture where grazing is restricted to two or three months each autumn.

Landscape value of saltbush. Saltbush can stabilize areas of bare saltland, reducing the risk of water and wind erosion and improving both soil structure and nutrient cycling. Saltbush can also prevent further rise in saline watertables [C.V. Malcolm, pers. comm.]. However in some areas, it may be necessary to incorporate deeper rooted salt tolerant trees in order to stabilize watertables.

Acacia (A. saligna)

Acacia saligna is indigenous to sandplain areas of the temperate zone of Western Australia.

Ironically this shrub is more widely used for forage and soil conservation in the Mediterranean region than in Australia [18, 41]. In the last five years however, some 5000 ha have been directly seeded into a range of soil types from deep sand to mildly saline and waterlogged sandy clays. Some of this is grazed by sheep.

Forage value of acacias. Acacias in general do not meet the forage requirements for maintenance of sheep. Dry matter digestibility is often below 50% due to a high lignin content and consequently animals on high acacia diets have difficulty reaching their daily energy requirement, and the build up of indigestible lignin can cause compaction. The digestibility of protein is also low due to the binding action of tannins [6, 27].

Being a leguminous shrub however they do have the advantage of maintaining higher protein levels in their foliage as it matures than grasses and other herbage. Crude protein contents varying from 14 to 19% have been reported from a range of different provinces of *A. saligna* from south-western Australia [R. Ferdowsian, pers. comm.].

Their role as a protein supplement to standing dry feed has not been confirmed by grazing experiments but would appear to be confirmed by two cases from commercial experience [30: A.N. Buchanan, pers. comm.].

In the second case, 2300 Merino ewes were maintained for five months on 60 ha of *A. saligna* and inter-row grass plus an adjacent 20 ha of dry annual pasture. The three year old plantation consisted of trees directly seeded into rows 10 m apart and was subdivided into five 12 ha blocks. The inter-row space was sown to a mixture of perennial veldt grass (*Erharta calycina* var. "Mission"), love grass (*Eragrostis curvula* var. "Consol") and serradella (*Ornithopus compressus*).

Newly shorn sheep were rotated through these blocks in one mob spending four weeks in each block. The acacia feed was further rationed by cutting only sufficient for six to seven days at one time. This assured a continual supply of the full range of foliage available, from young to old phyllodes.

The edible dry matter of tree fodder and grass were estimated at 4.0 and

2.0 tonnes per hectare, respectively. Sheep had constant access to the area of annual pasture (serradella and annual ryegrass) as this contained the only watering point. Feed available was estimated at 3 tonnes per hectare edible dry matter in January but there was no significant feed remaining in this area after two months.

The contribution of the acacia can be inferred from the fact that the estimated amount of grass and pasture would only be sufficient to maintain that number of adult sheep for about 90 days. In addition that feed would be very low in protein, probably around 4 to 5%.

Recent work with *A. aneura* in Queensland suggests it may be possible to increase the digestibility of protein by introducing exotic rumen flora to sheep in a similar way that toxicity in leucaena was overcome. Drenching sheep with rumen fluid from feral goats produced an increase in nitrogen digestibility of 12% and an increase in intake of 17% in one study [27].

Economic value of acacia. Studies that monitor liveweight, wool growth and vegetation changes are required before the value of acacia can be assessed in comparison with alternatives. In the above example, the saving in grain feeding at $1 per head per month is $11,500 or $144 per hectare of acacia plantation. In most years, the need for supplementary feeding would be only two to three months putting a value of $60 to $90 per hectare.

A distinct advantage of *A. saligna* is that it establishes readily from seed at a cost varying from $80 to $150 per hectare. A disadvantage of the feeding regime described above is the need for regular cutting. This was carried out by the farmer with a chainsaw, requiring six hours cutting per week. The current cost of mechanical pruning is $30 to $60 per hectare per year. *A. saligna* regrows vigorously when cut at 30 to 50 cm above ground level.

The expected life of a plantation under such management is unknown, although its survival in the wild is about 15 years.

Landscape value of acacia. The fact that Acacias establish relatively easily on a wide range of soils makes them a valuable addition to other species on degraded sites and other areas unprofitable under conventional agriculture.

On waterlogged and mildly saline sites, they can potentially increase draw down of saline watertables. On all sites they contribute protection from wind erosion and the ability to fix atmospheric nitrogen.

Acacia saligna is being used and studied as the basis for a system of wide spaced agroforestry on several farms where pasture or crop can be grown between the rows. In this case the forage value may be less important than the shelter effects of these rows, spaced 30 to 90 m apart (E.C. Lefroy, unpublished).

4.4. *Wet temperate zone*

This zone supports beef cattle and dairy cattle in addition to sheep for wool production. There are no large areas of fodder trees or shrubs in production although there are limited plantings of several species (see Table 3).

There is however increasing interest not only because of their ability to supply autumn feed but also because of the need to correct hydrological imbalance caused by the replacement of areas of native vegetation with annual pasture and crop.

The major disadvantage is the time necessary to protect trees from stock before they are productive. This is particularly so for the three species listed in Table 3 with edible fruit, which take 5 to 10 years to come into production. Yields of 100 kg per tree for mature honey locust and carob have been reported [8, 36] but there is no local data on animal response to these feeds to support their use.

Both the carob and the oak have been important in agroforestry systems of the mediterranean area providing forage as well as improvements in soil fertility and microclimate [15]. However, these systems have evolved over several hundred years and may not be relevant to the present economic and social structure of agriculture in Australia due to the labour intensive management required.

The willow and poplar species, however, can be managed as browse plants in much the same way as tagasaste and *Acacia saligna* [17]. Willows have been reported to have high protein and in vitro digestibility (16.2% and 65.8%, respectively) and to produce weight gains of 43 g per day in young sheep over a six week period as the sole feed [5].

Poplars such as *Populus deltoides* are used on a small scale in wide spaced agroforestry at densities of about 300 stems per hectare to provide timber and fodder from prunings [36].

5. Conclusion

The role of cultivated trees and shrubs as sources of fodder in Australia is that of supplying feed during periods of seasonal nutritional shortage. In northern Australia this period can last from April to October. In southern Australia it is generally from February to May.

Outside these periods, they can rarely compete with more abundant herbaceous feed due to their higher costs of establishment and management, their generally lower digestibility and the fact that they are often less accessible and acceptable to animals.

Only two species, *Leucaena leucocephala* and *Chamaecytisus palmensis*, have shown their ability to act as the sole source of feed for extended periods. The role of other species is in providing supplements of protein and other nutrients to dry pasture, and in the case of rangeland species, as the only source of nutrition during drought when all other feed is absent.

An important factor contributing to the commercial success of cultivated species is that they are easily and inexpensively established. The two species that have achieved commercial success are both leguminous shrubs that establish readily from seed.

A limitation with the present use of cultivated browse species is the dependence on stands of single species. The use of mixed species of trees and shrubs and the selection of complementary inter-row grass and herbaceous species would lessen the impact of insects and disease.

Management techniques developed to maximize production and persistence of cultivated browse plants involve short periods of intense grazing followed by long periods of recovery. Strategies similar to this may be necessary in the arid and semi-arid rangelands to retain palatable species that are now in decline.

References

1. Arieli A, Naim E, Benjamin RW and Pasternak D (1989). The effect of feeding saltbush and sodium chloride on energy metabolism in sheep. British Soc Anim Prod 49: 451—457
2. Chippendale GM (1968). A study of the diet of cattle in central Australia as determined by rumen samples. Tech Bull No 1, Primary Industries Branch, Northern Territory Administration, Darwin
3. Corbet HA (1951) Fodder trees; suggestions for their wider use. Imperial printing company, Perth
4. Craig GF, Bell DT and Atkins CA (1991) Nutritional characteristics of selected species of Acacia growing in naturally saline areas of Western Australia. Aust J Exp Agric 31: 341—345
5. Dann PR and Axelsen A (1986) Willow (*Salix* spp.) as drought feed for sheep. Proc Aust Soc Anim Prod 16: 408
6. Dann P and Low S (1988) Assessing the value of browse plants as alternative sources of fodder. Agricultural Science 1: 20—27
7. Dann P (1991) Fodder Trees in New South Wales. In: Proceedings, the Role of Trees in Sustainable Agriculture. Bureau of Rural Resources, Canberra
8. Esbenshade H and Wilson G (1986) Growing carobs in Australia. Goddard and Dobson, Melbourne
9. Everist SL (1969) The use of fodder trees and shrubs. Queensland Dept Primary Industry Advisory leaflet No 1024
10. Goodchild AV and McMeniman NP (1986) Nutritive value of Acacia foliage and pods for animal feeding. In: Turnbull JW, ed, Australian Acacias in Developing Countries. ACIAR, Canberra
11. Graetz RD (1986) A comparative study of sheep grazing a semi-arid saltbush pasture in two condition classes. Aust Rangel J 8(1): 46—50
12. Graetz RD and Wilson AD (1979). An assessment of herbivore diets in the chenopod shrublands. In: Graetz RD and Howes KMW, eds, Studies of the Australian Arid Zone. IV. Chenopod Shrublands, pp 144—159. CSIRO, Melbourne
13. Holm, AMcR (1986) The assessment of range trend in Western Australian pastoral shrublands. In: Rangelands: A Resource under Siege, Proc 2nd Inter Range Congr Adelaide
14. Imperial Agricultural Bureaux (1947) The use and misuse of shrubs and trees as fodder. Joint Publ 10, Imperial Agr Bureaux, Aberystwyth

15. Joffre R, Vacher J, de los Llanos C and Long G (1988) The dehesa: an agrosilvopastoral system of the Mediterranean region with special reference to the Sierra Morena of Spain. Agroforestry Systems 6: 71—96
16. Jones RJ (1979) The value of *Leucaena leucocephala* as feed for ruminants in the tropics. World Anim Rev 31: 13—23
17. Kraayenoord CWS and Hathaway RL (1986) Plant Materials Handbook for Soil Conservation. Soil conservation centre, Ministry of Works and Development, Wellington NZ
18. Le Houerou HN (1980) Browse in North Africa. In: Le Houerou HN, ed, Browse in Africa, pp 52—82. International livestock centre for Africa, Addis Ababa
19. Leigh JH and Mulham WE (1986) Selection of diet by sheep grazing semi-arid pastures on the riverine plain. Aust J Exp Agric Anim Sci 6: 468—474
20. Leigh JH, Wilson AD and Williams OB (1970) An assessment of the value of three perennial chenopodiaceous shrubs for wool production of sheep grazing semi-arid pastures. Proc XI Int Grasslands Congress 55: 9
21. Malcolm CV, Clarke AJ, D'Antuono MF and Swann TC (1988) Effects of plant spacing and soil condition on the growth of five *Atriplex* species. Agric Ecosystems and Environ 21: 265—279
22. Malcolm CV and Pol JE (1986) Grazing and management of saltland shrubs. J Dept Agric West Aust 27: 59—63
23. Malcolm CV, Swann TC and Ridings HT (1980) Niche-seeding for broadscale forage shrubs establishment on saline soils. Intern Sym on reclamation and management of salt affected soil, Karnal, India, Int Soc of Soil Sci 539—544
24. Marshall JK (1972) Principles of soil erosion and its prevention. In: Hall N, ed, The Use of Trees and Shrubs in the Dry Country of Australia. Aust Govt Pub Service, Canberra
25. Mattinson BC and Oldham CM (1989) Proc 16th Int Grasslands Congr Nice
26. McGowan AA (1991) The role of trees and shrubs as sources of fodder in Victoria. In: The Role of Trees in Sustainable Agriculture. Bureau of Rural Resources, Canberra
27. Miller S (1991) Improving the nutritional quality of mulga. The Mulga Line, Dept Primary Industries, Charleville, Qld
28. Mitchell AA and Wilcox DG (1988) Plants of the Arid Shrublands. University of Western Australia press with the Western Australia Dept of Agriculture
29. Morrissey JG and O'Connor REY (1988) Fair use and fair go; twenty-eight years of station management. Paper presented to the 5th Biennial Conference. Aust Rangeland Society, Longreach, Queensland
30. Nichols JE (1944) The behaviour of sheep browsing during drought in Western Australia. Proc British Soc Animal Prod, pp 66—73
31. Oldham CM, Allen G, Moore PM and Mattinson BC (1991) Animal Production from tagasaste. J West Aust Dept Agric Vol
32: 24—30 32. Pratchett D (1989) The prospects for leucaena on the Ord. J Dept Agric Western Australia 30: 2
33. Pratchett D (1991) Leucaena production on the Ord River Irrigation Area. Proc AMLRDC Shrub Legume Workshop, Brisbane
34. Purvis JR (1986) Nurture the land: my philosophies of pastoral management in central Australia. Aust Range J 8: 110—117
35. Quirk MF, Paton CJ and Bushell JJ (1990) Increasing the amount of leucaena on offer gives faster growth rates of grazing cattle in South East Queensland. Aust J Exp Agric 30: 51—4
36. Reid R and Wilson G (1985) Agroforestry in Australia and New Zealand. Goddard and Dobson, Melbourne
37 Salerian JS, Malcolm CV and Pol JE (1987) The economics of saltland agronomy. Technical Report 56, Div Res Man, Western Australia Dept Agric
38. Snook LC (1986) Tagasaste. Nightowl Publishers, Shepparton, Vic, 102 pp
39. Standing Committee on Agriculture (1990) Feeding standards for Australian Livestock: ruminants. Ruminant sub-committee, Standing Comm on Agric, Canberra

40. Stafford-Smith DM and Morton SR (1990) A framework for the ecology of arid Australia. J of Arid Environ 18: 255—278
41. Stringi L, Sarno R, Amato G, Leto G, Gristina L and Corrao A (1987) Shrub utilization in semi-arid environments: bio-agronomic aspects and their use in sheep husbandry. World Review of Animal Production XXIII (2): 79—90
42. Squires VR and Siebert BD (1983) Botanical and chemical components of the diet and liveweight change in cattle on semi-desert rangeland in Central Australia. Aust Rangel J 5 (1): 28—34
43. Squires VR (1989) Australia: distribution, characteristics and utilization of shrublands In: McKell CM, ed, The Biology and Uitilization of Shrubs. Academic Press, N.Y.
44. Torres F (1983) The role of woody perennials in animal agroforestry. Agroforestry Systems 1: 131—163
45. Vercoe TK (1986) Fodder potential of selected Australian tree species. In: Turnbull JW, ed, Australian Acacias in Developing Countries, pp 95—100. ACIAR Proceedings No 16 ACIAR, Canberra
46. Warren BE, Bunny CJ and Bryant LB (1990) A preliminary examination of the nutritive value of four saltbush (*Atriplex*) species. Proc Aust Soc Anim Prod 18: 424—427
47. Watson IW and Holm AMcR (1990) The benefits of palatable shrubs for wool production in a semi-arid environment. Proc 6th Aust Range Conf Carnarvon
48. Wesley-Smith R (1991) The use of woody vegetation for fodder in the Northern Territory in The role of trees in sustainable agriculture. Bureau of Rural Resources, Canberra
49. Wheeler JL and Freer M (1987) In: Alexander and Williams, eds, Pasture and Forage: The Feed Base for Pastoral Industries, pp 164—182. The Pastoral Industries of Australia
50. Wilcox DG (1979) The contribution of the shrub component in arid pastures to the production from sheep. In: Graetz RD, ed, Studies of the Australian Arid Zone. IV, pp 170—177 Chenopod Shrublands. CSIRO, Melbourne
51. Wildin JH (1991) The role of trees as a source of fodder in Queensland. In: The Role of Trees in Sustainable Agriculture. Bureau of Rural Resources, Canberra
52. Wilson AD (1969) A review of browse in the nutrition of grazing animals. J Range Man 22: 23—28
53. Wilson AD (1977) The digestibility and voluntary intake of the leaves of trees and shrubs by sheep and goats. Aust J Agric Res 28: 501—508
54. Wilson AD (1990) The effect of grazing on Australian ecosystems in Proc Ecol Soc Aust 16: 235—244

Agroforestry Systems **20**: 141–166, 1992.

Retention of native woody vegetation on farms in Australia: management considerations, planning guidelines and information gaps

J. C. SCANLAN[1], R. PRINSLEY[2], J. P. PIGOTT[3], S. WAKEFIELD[4], F. VAN DER SOMMEN[5], F. DUNCAN[6], T. STADLER[6], R. McLELLAN[7] and A. FARAGO[7]

[1] *Department of Primary Industries, Charters Towers QLD, Australia;* [2] *Rural Industries Research and Development Corporation, Barton ACT, Australia;* [3] *Conservation and Land Management, Narrogin WA, Australia;* [4] *Department of Agriculture and Fisheries, Bathurst NSW, Australia;* [5] *Conservation Commission of the Northern Territory, Darwin NT, Australia;* [6] *Forestry Commission, Hobart TAS, Australia;* [7] *Department of Conservation and Environment, Melbourne VIC, Australia*

Key words: land degradation, fragmentation, configuration, legislation, economics

Abstract. Australia's forests and woodlands have been extensively cleared since European settlement. Cropping regions have the least retained native vegetation, with the greatest depletion of forests and woodlands being observed in Western Australia and South Australia. Salinisation, erosion and landslides have resulted from excessive clearing and inappropriate tree removal from particular units within the landscape. Major problems in cropping regions are the dramatic reduction in total woody cover and the fragmentation of retained vegetation into isolated remnant patches which tend to be very small (< 10–15 ha). Those states with little remnant vegetation have legislation which restricts further clearing and the main emphasis is on managing remnant patches for conservation. For those states with a large proportion of woody vegetation uncleared, the main priority is to ensure that past mistakes are not repeated and that any vegetation clearance is based upon sound ecological principles.

Throughout Australia, the condition of the remaining vegetation is of great concern and management guidelines for both conservation and production are being sought. Clearing may lead directly to soil erosion, but often the cause of erosion is inappropriate post-clearing management. Overgrazing is a major cause of vegetation and soil degradation, particularly in semi-arid regions.

An understanding of the complex inter-relationship between woody vegetation, grazing and fire is necessary if 1) flora and fauna are to be conserved; 2) hydrological balance is to be maintained; and 3) rural industries are to remain viable. Planning at the catchment level is necessary to implement strategies to meet conservation and hydrological goals. Within that framework, individual property planning must include consideration of the area of native woody vegetation to be retained and its configuration. In many cases, the purpose of retention will influence the decisions of where and how native vegetation should be retained.

1. Introduction

Before European settlement, forests and woodlands covered 33% of Australia's land area. An estimated 50% of forests and 35% of woodlands have now been cleared or severely modified [19]. The resultant decline in animal and plant populations over the last two centuries has been marked with 78 species of plants and 17 species of mammals thought to be extinct [50]. Much of the land degradation in southern Australia is a consequence of excessive clearing of vegetation. More than half of grazing and cropping lands — over 2.6 million square kilometres — may require treatment for forms of degradation [19]. In Western Australia alone, losses associated with land degradation are estimated to be $600 million annually.

Thus, the economic benefits of clearing native woody vegetation in Australia have incurred substantial costs. Clearing continues in some states, despite the knowledge that retaining woody vegetation on farms has advantages including direct economic benefits (timber, honey production, shelter for stock), indirect benefits (maintenance of soil fertility), off-site benefits (maintenance of water quality in rivers) and natural heritage value. This occurs because landholders often see retention of native vegetation as an unacceptable cost in foregone production.

The challenge for Australia is to meet multiple land use requirements and to reach compromises where there is conflict between the requirements for different uses. The concept of sustainable agriculture requires a basic change in attitude of many landholders and agricultural scientists. The aim of maximising production and optimising income in the short term must be accompanied by an objective to achieve sustained yields over the long term by the maintenance of resources.

Landholders make a living from rural resources. Action to conserve the native vegetation resource may incur direct costs (e.g. fencing) and these costs must be balanced by some economic benefit if the business enterprise is to survive. To date, conservationists have not paid sufficient attention to documenting the economic benefits of native vegetation retention, especially in relation to crop and animal production. It is imperative that the person who makes a decision to clear vegetation does so only after collecting and evaluating all available information on the full impact of any clearing. However, if technical specialists (conservationists, ecologists, agronomists) cannot provide objective, quantitative information, the landholder cannot be blamed for making inappropriate decisions.

2. National consideration of vegetation retention

The Australian community generally wishes to ensure that their standard of living is maintained through profitable agriculture and other industries, and accepts that modification of the original vegetation has been necessary to achieve current levels of agricultural production. However, there is concern about the sustainability of current practices and these have been the subject of a recent federal parliamentary report [25] which makes 27 recommendations on sustainable agriculture.

2.1. *Extent of clearing*

Under the Australian constitution, states have the responsibility for protecting and regulating natural resources. Differences exist between the states in the extent of clearing and in the regulation of land use, with climate, history of development and vegetation resources dictating the major variations. Compared to the other states, for instance, Queensland, Northern Territory and northern Western Australia have a large proportion of native vegetation remaining intact (see Prinsley, [48] in this issue for location map). In these states, grazing by sheep and cattle is the main form of land use and the emphasis is on ensuring any vegetation clearance is based on sound ecological principles. By contrast, emphasis is on retention and management (conservation) of the limited area of remaining native vegetation in most agricultural (cropping) zones.

The degree of clearing of native woody vegetation communities varies greatly across Australia (Fig. 1 [5]). The area of grassland has doubled, and small increases in low open woodlands and tall open shrublands have been recorded since European settlement. These changes have resulted from clearing other vegetation communities. However not all undesirable changes in Australia's vegetation have been associated with removal of native vegetation. Increased densities of native woody species in *Acacia aneura* and *Eucalyptus populnea* woodlands in eastern Australia have been significant

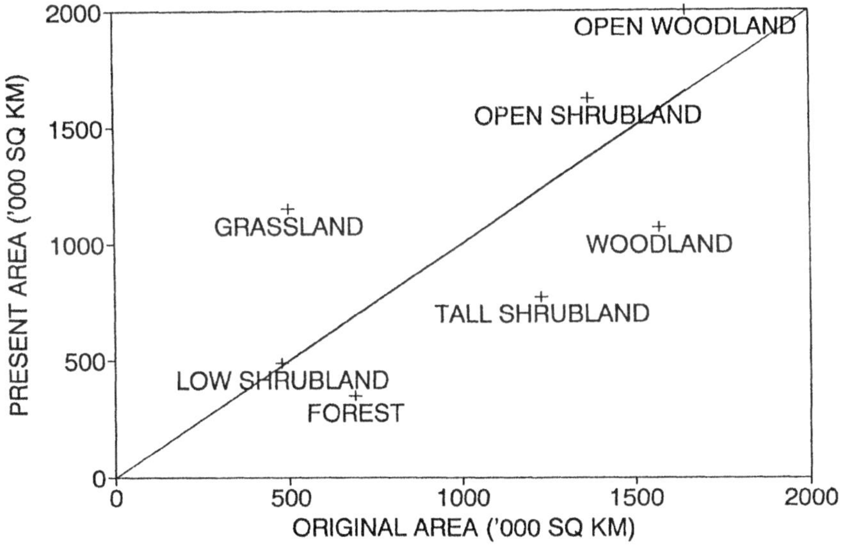

Fig. 1. The current area of vegetation types in Australia (1980s) compared with the estimated area in the 1780s [from 5].

[27] with inedible species of *Eremophila, Cassia* and *Dodonaea* causing decreased pasture production and increased mustering and management costs. The introduced *A. nilotica* is present in 6 million ha of northern mitchell grasslands [13] and these productive grasslands are in the process of becoming thorn shrublands of little value for grazing. Other introduced trees and shrubs are also increasing in area and density in northern Australia [62].

The degree of clearing and some associated impacts are shown in Table 1. Some examples of associated degradation are highlighted in the following section. Detailed discussions of the situation in each state are provided in the proceedings of the conference on 'Role of Trees in Sustainable Agriculture' [2].

Table 1. The extent and consequences of clearing within each state of Australia. (Data drawn from [2] where not shown).

State	Extent of clearing	Major forms of degradation
Queensland	~ 16% of state [10]; mainly on freehold land	native and exotic woody weeds are serious [62]
Northern Territory	< 2% of state cleared	potential exotic woody weed threat [62]
Western Australia	88% of cropping zone cleared; < 1% of pastoral zone cleared; 33% is unoccupied	widespread salinisation of cropland in south-west [63]; fragmentation of remnant vegetation serious in cropland; 5 of 11 macropod species locally extinct in south-west [4]
South Australia	84% of cropping zone cleared; < 1% of pastoral zone	27% of land mammals extinct [26]; 33 plant taxa extinct [35]; fragmentation serious in cropland
Victoria	65% of native forests cleared	25% of flora extinct, rare or endangered; 33% of vertebrates rare or threatened; 3.2 M ha eroded, acidified or compacted; 7.8 M ha at risk of degradation
Tasmania	24% cleared	~ 20% of plants poorly represented in conservation areas; 18% of state shows some form of land degradation [51]
New South Wales	40% of forests and woodlands cleared	7% of eastern and central divisions seriously eroded; 70% of western division with native woody weed problem

2.2. *Adverse impacts of clearing*

2.2.1. *Land degradation*

Eighteen percent of Tasmanian land has suffered some form of land degradation [51], much of this associated with tree clearing. While there is general recognition of degradation problems at state or regional level, few landholders accept that problems exist on their own properties [53] and even fewer associate economic losses with the loss of farm trees [40].

Clearing of native woody vegetation for agriculture is accepted as the primary cause of land degradation in the agricultural regions of South Australia and Western Australia. Dryland salinity, waterlogging, water erosion, wind erosion, water repellency, soil acidification, soil structure decline, subsoil compaction, decline of remnant native vegetation, management of the conservation estate, rehabilitation in the mining industry and degradation of waterways are prominent issues [72, 73]. Soil acidification has great potential for land degradation and is predicted to affect 10.5 million ha, or 55% of the agricultural agroecosystem, over the next 100 years. Overall, the annual loss of production is estimated to be $600 million.

Less than 1% of the pastoral zone of Western Australia has been cleared but pasture degradation has been extensive e.g. up to 40% of the Nullarbor region has been degraded [42]. Land degradation in this state's grazing lands is estimated to cost $37 million per year with the greatest problems being experienced on more productive and better watered parts of the zone [46].

In Victoria, recent data on land degradation are summarised in the draft 'Decade of LandCare Plan' [69] and include:

- 240,000 ha of irrigated land are seriously salinised (likely to increase four-fold over 50 years whether or not remedial action is taken);
- 3,200,000 ha of land are affected by erosion, acidity or compaction;
- 7,800,000 ha are at high risk of degradation;
- 65% of streamlength in cleared areas is in poor condition.

2.2.2. *Salinisation*

Dryland salinity problems are widespread in Western Australia and Victoria as a result of excessive tree clearing (see Schofield [63] in this issue). The risk of dryland salinisation may not be as severe in northern Australia as in southern states [9]. The major argument for this view is that in the Mediterranean climate of southern Australia, rainfall during the winter growing season often exceeds evapotranspiration, resulting in deep percolation and mobilisation of incipient salt, whereas in Queensland, very high rates of evapotranspiration occur during the summer growing season. However, monsoonal regions of Australia can also be characterised by substantial deep drainage and leaching of salts [74] and recent work in subtropical *Acacia harpophylla* forests also suggests increased deep drainage and movement of salt as a result of tree clearing [66].

2.2.3. *Woody weeds*
Increased density of native woody species in semi-arid woodlands causes economic hardship for graziers in northern and eastern Australia. Regrowth from native *Eucalyptus* and *Acacia* is also an economic problem for landholders in the eucalypt woodlands of eastern Queensland [10, 58]. A majority of the *Eucalyptus populnea* and *Acacia aneura* communities of New South Wales and Queensland is affected to some degree by increased density of native woody species (*Eremophila, Cassia, Dodonaea*). This leads to reduced lambing percentages, reduced wool cut per head, reduced carrying capacity and increased stock handling costs [8]. Overgrazing and soil erosion also result from the proliferation of these native species.

The introduced exotic tree, *Acacia nilotica*, is spreading rapidly through the northern mitchell grass (*Astrebla lappacea*) zone in Queensland which is one of the world's most extensive and productive natural grasslands. At present some 6 million ha of the grassland contains at least some of this woody weed [11] and these areas are likely to be transformed into thorn shrublands. The introduced woody climber *Cryptostegia grandiflora* is widespread in the woodlands of central and northern Queensland [14]. Riparian zones become completely covered with this plant with severe reductions in the value of these areas for domestic livestock production and wildlife habitat. Other woody weeds in pasture lands of northern Australia also interfere with livestock production [62].

2.2.4. *Fragmentation*
In the main cropping regions of Australia, the proportion of remaining native vegetation is very small. Added to this is the fragmented nature of this remaining vegetation. In a survey of remnant vegetation in the Midlands Division of Western Australia, two-thirds of patches were less than 16 ha in size and only 8% exceeded 50 ha [4]. Only 60% of the 53 patches were regarded as relatively undisturbed. Many of these remnants had high edge to area ratios [57], leading to increased probability of ingress of flora and fauna of neighbouring vegetation types, and to increased fire risk. Associated with this clearing and fragmentation, 5 of 11 macropods present at the time of European settlement are locally extinct and only 2 are still common [4]. Local extinction of bird species [55] and invasion of woodlands by birds of the pastoral regions has also occurred [56]. Decreased water quality has been observed where more than 50% of catchments in the south-west has been cleared [64].

A study of the extent of fragmentation of native vegetation on the Fleurieu Peninsula of South Australia revealed that only 9% of the region was carrying original vegetation and 67% of this was located in blocks less than 10 ha in size [75]. In an area of 150,000 ha, there were only 3 patches in excess of 500 ha.

2.3. *Greenhouse gas emissions*

Global climatic changes may result from increased atmospheric concentrations of radiatively-active ('greenhouse') gases such as carbon dioxide, methane and nitrous oxide. The gases can be directly linked to agricultural development, e.g. methane is emitted by herbivores and termites; burning releases carbon dioxide, carbon monoxide, methane, and various oxides of nitrogen (see [44, 20] for a general discussion of agricultural contributions to greenhouse gas emissions and global warming). The proportion of fixed carbon that becomes organic matter with a long half-life, and the proportion that is decomposed directly versus that consumed by herbivores have a marked impact on the greenhouse gas emissions. Forests and woodlands contain a large bank of carbon sequestered in woody material and a large proportion of this is released into the atmosphere following tree clearing. Since white settlement, land clearing in Australia has resulted in the loss of 7 Gt of carbon (14 Gt of standing biomass) (Gt = 1,000 million tonnes), with the present standing biomass containing an estimated 20 Gt of C (calculated from vegetation areas [5] and biomass estimates [6]). This represents a contribution of 3—5% to the world's increase in CO_2 within the atmosphere.

Methane is more effective as a greenhouse gas than CO_2. Increased grazing animal populations are one source of increased levels of this gas in the atmosphere. Minor changes in pasture and stock management can result in significant decreases to total greenhouse gas emissions, without a commensurate decrease in farm income [31]. A 20% reduction in greenhouse gas emissions from grazed pasture systems in northern Australia could be achieved by lowering stocking rates with little economic cost in terms of lost production [31]. An added benefit would be reduced soil erosion resulting from greater ground cover [23].

3. Role of legislation in management of native woody vegetation

The way to achieve sustainable use of native vegetation resource is not clear. Regulation can result in reduced environmental damage by forcing particular forms and levels of activity. Such regulations can be politically unpalatable as well as hard to enforce, slow and difficult to adapt to change, and their consequences may be off-target. Possible alternatives include price-based measures (charges and subsidies) and rights-based measures (right to use or abuse). Although largely untested in Australia, market-based measures may increase the relative returns from environmentally benign options compared with management or processes that cause damage to the environment [54]. Such measures may be easier to fine-tune than direct regulations, and may allow a smoother integration of economic and environmental considerations. However, market forces do not necessarily reflect ecological stress, and biological systems may pass an irreversible threshold before the economic impacts force a management change [54]. Subsidies may actually increase resource degradation by reducing the cost of production and encouraging overuse of resources [41].

At present, legislation governing native woody vegetation varies quite considerably between the states. There are 33 Acts of Parliament within the states and territories that are partially or wholly concerned with vegetation management, with ten separate Acts in New South Wales alone. These range from voluntary conservation schemes to stringent regulations which effectively prevent large scale clearing of native vegetation (e.g. in South Australia). Many farmers have short-term financial pressures which force them to clear native vegetation. Schemes offering financial assistance (18 of the 33 Acts have some financial support for landholders) may enable some of these landholders to refrain from clearing. However, voluntary conservation schemes and financial incentives may not provide sufficient protection where conservation is a high priority (e.g. to protect an endangered species). In these cases, legislation may be necessary.

A relatively small proportion of northern Australia (Queensland, Northern Territory, and northern half of Western Australia) has been cleared (Table 1) and clearing legislation is not as stringent as in southern Australia. However, these state governments are concerned about adverse environmental aspects of excessive tree clearing. The procedure for applying for permits to clear trees on leasehold land within Queensland is under review. One proposal is that a whole-farm management plan must accompany any clearing application, and the Queensland Department of Primary Industries and Queensland National Parks and Wildlife Service should be involved in assessing applications [78]. Consideration is also given to the possibility of controlling clearing on freehold land.

In Western Australia, new soil conservation conditions require fencing of all newly cleared land. The maximum clearing allowed is now viewed in the context of the remnant vegetation within the whole catchment rather than simply the amount remaining on any particular farm. This can lead to conflict where landholders need to clear a high percentage of their farm to remain profitable in the short term.

Clearing native vegetation in Victoria is controlled by a number of Acts and regulations [39], including the Planning and Environment Act of 1987. This act and the sudden temporary clearing controls in 1989 were reinforced by an amendment which provided permanent clearing controls. On introducing the controls, the Victorian Government adopted the policy that there should be no further broadscale clearing in Victoria. Introduction of these controls has already been effective in substantially slowing the rate of clearing [39].

A system of Heritage Agreements was introduced in South Australia in 1980 to protect native vegetation and to encourage landholders to manage native vegetation for conservation. From 1980 to 1985, this voluntary scheme resulted in 10,500 ha being covered by 125 agreements. Landholders receive rate relief, and the state government erects all fences and assists with additional costs above those required by various acts. In 1985, a stronger legislative and economic base for the Native Vegetation Management Act was introduced. Since then, an additional 400 landholders have entered into Heritage Agreements, protecting some 270,000 ha in the agricultural region. The scheme is expected eventually to protect 50% of the remaining native vegetation in the agricultural areas outside the state government reserve system.

4. Economic aspects of management of native woody vegetation

4.1. *Valuing natural resources*

Insufficient monetary value is placed on Australia's natural environment by the community [25], and whenever resources are undervalued or free, they are overused or abused. There is a need to put an appropriate value on resources, and to define clearly the rights and responsibilities of landholders and the wider community to ensure that resources are used in a sustainable manner. Many of the proposed solutions, including legislation and subsidies, can be criticised from an economic viewpoint [34].

The value placed on a natural resource can have a significant effect on its use. In Queensland, the Brigalow Development Scheme of the 1970s resulted in the formation of hundreds of new grazing properties based on the clearing of *Acacia harpophylla* (brigalow) woodlands and their replacement with perennial grass pastures (based on *Cenchrus ciliaris*). These lands were allocated by ballot with a major requirement being that a minimum area had to be cleared within a certain time. This resulted in some inappropriate clearing being undertaken and some marginal lands were also cleared and 'developed'. Current agricultural production (beef and grain) in the region is vastly in excess of the meagre beef production possible under the original woodlands, but there has been a substantial environmental cost. The scheme was so successful that very few stands of uncleared brigalow remain intact. Because the heavy clay soils have high pH and high soluble salts at depth, there is increasing concern that these lands may suffer salinisation problems in the future and that stream and groundwater quality may be adversely affected [66]. Thus, costs to the community of additional clearing may include expenditure to prevent or ameliorate degradation of land and water resources.

4.2. *Economic benefits of native vegetation*

Managed native forests offer farmers income diversity, however the low financial inputs may be mirrored in the modest income generated by most native forest production [45]. In an evaluation of alternative uses of blackwood swamp forest in north-west Tasmania, management for production of *Acacia melanoxylon* (high value cabinet timber) was compared with clearing and development of pastures for dairying. Forestry was shown to be more profitable if the cost of capital was high (20%) while dairying was superior at a rate of 15% interest [33]. This analysis was based on timber royalties of $AUS15 m^{-3}, but royalties are currently $40 m^{-3} and may rise to >$100 m^{-3}. Such increases in the value of timber products are likely to be common and will make forestry much more attractive economically in the future.

In 1971, the advent of a Tasmanian woodchip export industry stimulated logging of native forests and 40—50% of cleared forests were converted to agricultural usage. In many situations, the money from timber sales was used for subsequent pasture development. Only 36% of the logged areas were actively replanted with trees [65]. Over the next 40 years, most private land supporting native forests will be sold at least once, potentially making these areas available for timber production [65]. There is strong landholder support to lift the current ceiling of 800,000 tonnes/year of woodchip from private land [21, 49] to enable logging of old-growth forests while a secure market exists. Under the proposed Tasmanian Forest and Forest Industry Strategy [22] the ceiling will only be lifted once a mechanism to protect environmental values on private land has been developed.

Tree clearing may not be the most economic form of development in the eucalypt woodlands of north Queensland [24] with legume augmentation of native pastures being more viable and sustainable. Legume augmentation with *Stylosanthes scabra* does not require any tree removal nor does it require any destocking during establishment. Water use by the shrubby *S. scabra* resembles the pattern of tree water use rather than that of grass water use [74]. Such a pasture system (perennial stylo in an undisturbed *Eucalyptus* woodland) will cause little disruption to the regional hydrological balance and will not lead to rising water tables as is possible following tree removal.

4.3. *Cost of retaining native vegetation*

Costs of retaining native vegetation fall into three categories: 1) direct costs associated with protecting remnant vegetation, e.g. fencing and other management required to prevent grazing, fire damage and weed invasion; 2) potential production may be foregone and this may influence the viability of a particular property; 3) the density of native species may increase to such an extent that production is severely restricted and threatens property viability.

The direct cost of the Native Vegetation Retention Program in South Australia has been $40 million since 1985. This is by way of compensation for 400 landholders being refused permission to clear land on their properties on biological grounds and where the landholder has entered a Heritage Agreement. A further $1—2 million per year will be spent on the management of these lands. Despite the high initial outlay, the program has been economical in terms of the cost per tree retained, being about 15c per tree

[67]. This compares with $2—10 for tree seedlings planted and about 19c per tree for direct seeding costs.

Loss of potential production is seen as a severe impact of not clearing native vegetation. In Victoria, clearing at the rate which occurred up to 1987 added less than 0.1% per annum to the area of already-cleared land. On a statewide basis, this is an insignificant contribution to economic activity. Fluctuations in markets for agricultural commodities such as wheat and wool alone cause much greater changes in the gross value of agricultural production.

Increased density of native shrubs and trees is common in the majority of the western division of New South Wales. Reduced lambing percentages, reduced wool cut per head, reduced carrying capacity and increased stock handling costs [8] adversely affects property viability in the region. The estimated gross margin for land free of unwanted woody plants is $30 ha^{-1} compared with $13.50 for land with high densities of woody species.

5. Vegetation management at the property level

5.1. *Property planning*

A property plan should form the basis for any major change to the management of a property, particularly when clearing or replanting native woody vegetation is concerned. This has not been common practice in the past. Currently, there is a move to encourage all landholders to develop detailed property plans which not only include the physical resources but also enable management changes to be identified and implemented [25]. Where the clearing of native vegetation is concerned, state departments either require or recommend the production of a property plan to ensure a balance between development, the owner's physical and financial resources, and environmental considerations.

5.2. *Retention of remnant vegetation*

5.2.1. *Area of remnant vegetation*

About 5% of Australia is protected by World Heritage Agreements or parks and reserves [25]. There are large areas in Western Australia, Northern Territory and South Australia that are either vacant crown land or are under leases that involve low levels of land use, particularly traditional aboriginal land use. The problem in Australia is not that there is insufficient total area of undisturbed vegetation, but rather that the distribution of this vegetation is non-uniform and that many vegetation types are grossly under-represented in the reserves system. The arid parts of the continent are uncleared but have been disturbed by the grazing of sheep, cattle and feral animals. Thus native vegetation has not been removed but it has been markedly disturbed. Overgrazing, herbaceous vegetation change and subsequent soil erosion are major issues in these lands, as is the increase in undesirable native woody plants [27]. In contrast to the arid zone, areas suitable for agricultural crop production and to a lesser extent for pasture development have undergone the greatest degree of clearing. Also, naturally treeless areas with fertile clay soils (e.g. the Darling Downs and Central Highlands in Queensland) have been cropped to such an extent that few areas of natural grasslands still exist.

Two separate considerations relate to appropriate retention of native vegetation. The first is the total area that should be retained, and the second is configuration (spatial arrangement) of retained vegetation. The total area to be retained may depend on regional hydrological, wildlife and soil movement considerations. As discussed in Section 4, it is difficult to determine an appropriate resource value for native vegetation. Extensive tree clearing has been associated with the relative ease of showing some economic benefit from tree clearing (at least in the shorter term of c. 10—15 years). Once the total amount of vegetation to be retained has been determined, the aim must be to maintain the maximum degree of biodiversity possible. The habitat size required for, and appropriate population sizes of flora and fauna are not well understood for even the most common species, making judgements about the required size of reserves more subjective than desirable.

5.2.2. *Fragmentation of remaining vegetation*

A consequence of agricultural development has been the vast reduction in native vegetation remaining intact, and the high degree of fragmentation that exists within the remaining vegetation [71]. Fragmentation leads to two problems. Firstly, the viability of isolated small blocks of vegetation is generally poor [29], even when these are protected from use by landholders and their domestic livestock. Conservation of flora may be achieved by reserving such areas, but fauna is at risk. In South Australia, 27% of land mammals have become extinct [26] whereas only 6% of plant taxa have become extinct in the state [35]. The second aspect of the problem is lack of habitat diversity. In well developed agricultural areas, remaining vegetation tends to be a non-random sample of former habitats: swamps that could not be drained; steep hillsides; small patches of rainforest maintained for 'conservation'. These remnant patches support a narrower spectrum of flora and fauna than if all habitats were represented. Isolated blocks are not efficient in supporting a wide variety of native fauna and are prone to degradation from salinity, grazing, weed invasion and rising water tables [28]. Interconnecting these remnant areas with wildlife corridors may enhance the value of reserved areas. A framework for the re-integration of remnant vegetation patches has been developed [30].

Reserves must be large enough to encompass the spatial processes and ecological gradients of the system being conserved [68]. Arguments can be proposed in favour of several small reserves as opposed to one large reserve. Those in favour of many small reserves concentrate on the number of species in these reserves and on which particular species can be protected (e.g. endangered species). A network of smaller reserves contain different (and more) species than a single large reserve of equivalent total area [37]. Arguments favouring large reserves focus on the lower probability of extinction in large populations. A mixed collection of small reserves and large areas may be the best policy as some species are best protected by small reserves and others by large ones. Obtaining an answer to the question of the appropriate reserve size involves studying the biology and habitat needs of individual species, their minimum area requirements as well as factors restricting species distribution [1]. Explicit procedures for selecting reserves for a nature conservation network have been proposed [36, 37].

5.2.3. *Configuration of remnant vegetation*

Once a decision to retain a given area of native vegetation has been made, the next issue is where to leave that vegetation. The trend has been to clear the most productive elements within a landscape completely and to leave surrounding, less productive areas intact. While this may ensure that a large area of the property or paddock remains covered in woody vegetation, it reduces habitat diversity. Therefore any clearing should not exceed a maximum proportion of each habitat type. That proportion will vary between regions and the appropriate values to meet the multiple goals of agricultural production, water quality and fauna and flora conservation are unknown for most regions.

The most appropriate configuration of retained vegetation will depend on the planned land use. Where potential salinisation is a concern, intake areas should support native woody vegetation; where shelter from cold winds is required, south and south-western slopes should be left covered by trees; where trees actually encourage growth beneath their canopy, some scattering of trees may be beneficial. Scattered trees may also reduce temperature extremes and in tropical to subtropical environments this may be sufficient to prevent frosting of pastures [J.G. McIvor, pers. comm.].

Retained vegetation can be in the form of strips or clumps. Clumps are more effective in that a greater proportion of the total area is natural (no edge effect) compared with strips (Fig. 2). The wider the strip, the larger the natural habitat area for the same reserved area. Clumps must be large enough to support a viable population of wildlife and should be joined to other native vegetation areas to allow movement between major reserved areas.

A disadvantage of a high proportion of edge habitat is that this transition zone is prone to degradation and is not representative of the central zone of clumps or strips. The edge zones support some flora and fauna typical of the adjacent vegetation type. In cases where the retained vegetation borders sown pasture, plant invasion can be a significant form of natural habitat degradation. One sown pasture species, *Cenchrus ciliaris*, has been regarded by some groups to be one of Australia's top environmental weeds [32], even though it is one of the most widely planted pasture species in Queensland

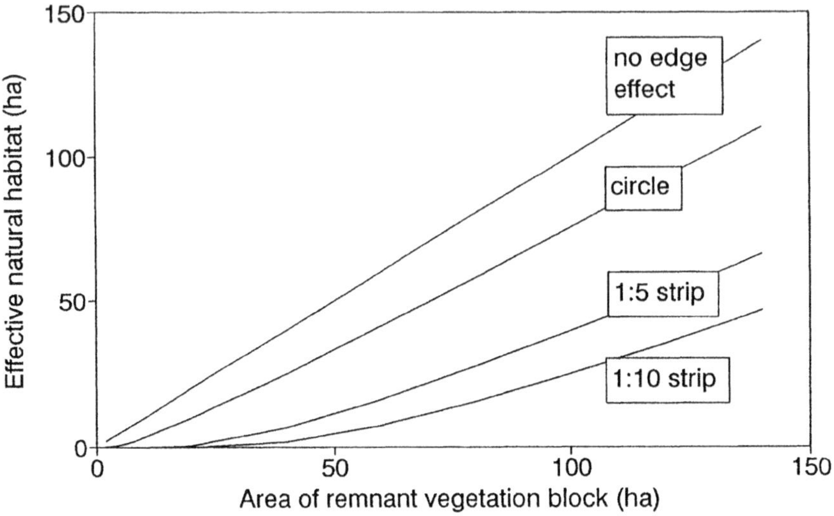

Fig. 2. The effect of shape of retained vegetation on the area of natural habitat with that vegetation. Natural habitat is defined as that part of the block greater than 75 m from the nearest boundary. The 1:5 strip is 5 times as long as wide, and 1:10 strip is 10 times long as wide. The difference between the 1:1 line (no edge effect) and the particular configuration line is an indicator of the area of 'edge' habitat (from calculations by JCS).

[70]. The weed potential comes from its ability to spread, especially along watercourses in the arid and semi-arid interior. The presence of *C. ciliaris* increases local stocking pressure and many native herbaceous species are depleted by the combination of direct competition with *C. ciliaris* and grazing.

Another option for retaining some woody vegetation and increasing livestock carrying capacity is to thin the existing stand, leaving a savanna landscape. This has aesthetic appeal, but in most situations a savanna is undesirable. Problems include 1) that the habitat for native fauna and flora is dramatically altered; 2) that remaining trees are likely to have a shorter lifespan, and are more prone to insect and disease attack, and to fire damage; 3) that pasture production is often less from a savanna landscape than from an area in which the equivalent number of trees was retained in undisturbed habitat while the complement of the area was cleared [10]; 4) the mature remaining trees are seed trees that ensure a seed source to re-establish new trees, thereby creating an inherent regrowth problem. As few as 40 trees/ha, each 10 m high, can result in three quarters of an area being subject to seed rain (Fig. 3), whereas the same number of smaller trees potentially impact a mere 10% of the area.

A savanna landscape can be desirable where the canopied zone is more

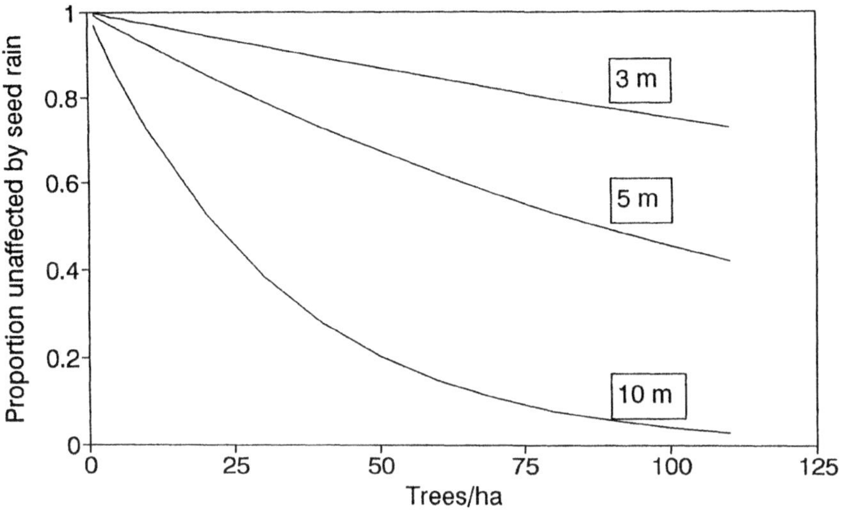

Fig. 3. The proportion of an area that is not subject to 'seed rain' for a range of plant heights and densities. Seed is assumed to fall within a distance equal to plant height: particular species will vary widely from this rule but the principles remain (from calculations by JCS).

productive than the interspaces. Reasons for this include increased total soil nutrient levels [17, 18], and indirect effects of shade [76, 77].

5.3. *Soil erosion*

Clearing woody vegetation is often equated with land degradation and soil erosion. While this is certainly true where disturbance leads to the exposure of a dispersive B horizon in the soil, it is not a necessary consequence. In most cases, it is post-clearing management of grazing that leads to soil erosion in grazing lands rather than the act of clearing per se. This separation of the causes of damage into direct effects of clearing and post-clearing management is necessary if degradation is to be avoided or minimised.

Heavy grazing may cause sheet erosion irrespective of the presence of trees [23] as soil erosion in pasture land is greatly influenced by the extent of surface cover of vegetation. Densely timbered areas may have either more or less total cover of understorey plants and tree leaf litter than a cleared site depending on the species involved (Fig. 4). Trees compete directly with grass for water and nutrients and, in all but the monsoonal zones, this usually results in less herbaceous cover under trees than in cleared areas [43]. Tree litter complicates this effect. This is exemplified by comparing Figs. 4a and 4b. In *Eucalyptus populnea* woodlands (Fig. 4a), there is a decline in total ground biomass (tree leaf litter plus pasture) as tree density increases due to the overriding negative effect of increasing tree density on grass cover. In

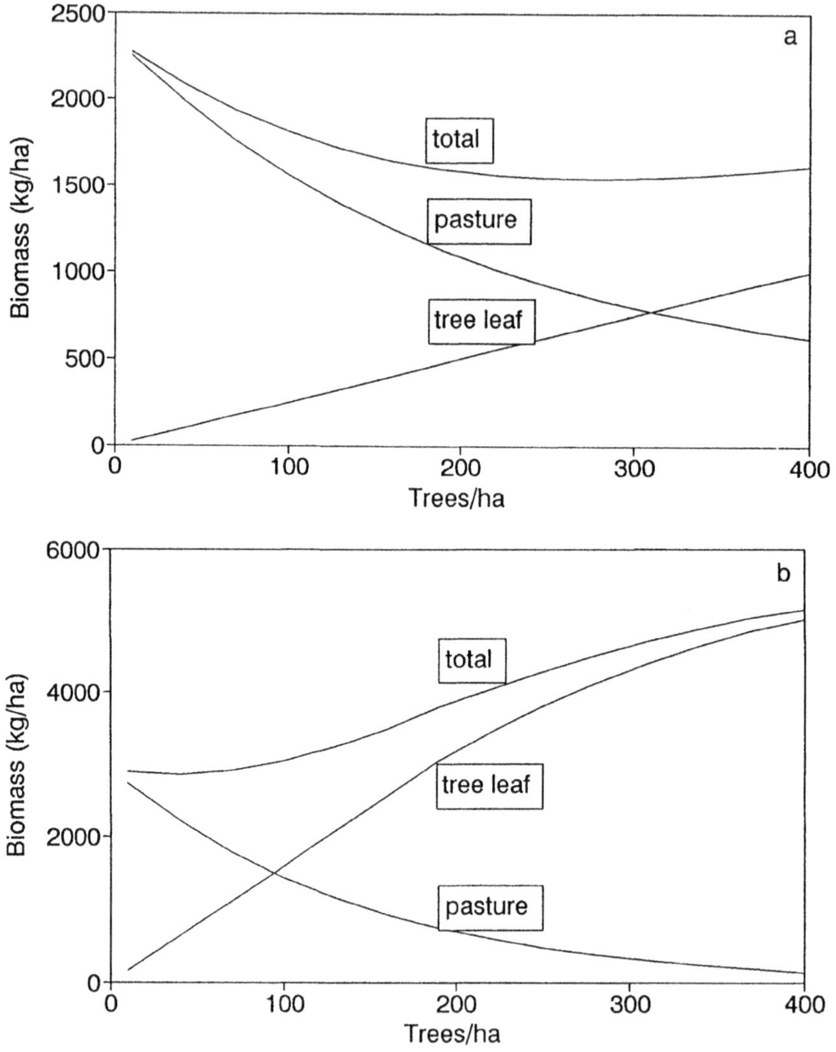

Fig. 4. Biomass of tree leaf and pasture within (a) *Eucalyptus populnea* and (b) *Acacia harpophylla* communities in Queensland. Data for *E. populnea* from [11] and for *A. harpophylla* from [59].

Acacia harpophylla communities (Fig. 4b), however, the highest total ground biomass is at the highest tree densities due to the higher relative production of tree litter. Ground cover is further modified by the interaction of grazing pressure, tree cover, use of fire, and rainfall amount, intensity and distribution. In the Northern Territory, these factors combine to produce less surface runoff from natural woodlands than from cleared pasture [16]. This contrasts with north Queensland where much greater runoff was reported from natural grazed woodlands than from cleared native pastures [23]. The interaction of grazing pressure, tree cover, use of fire and the amount, intensity and distribution of rainfall creates a complex set of erosion responses. Therefore generalisations about tree cover and surface soil erosion are not possible.

Clearing initially exposes the soil surface to higher surface temperatures which break down organic matter and reduce aggregate stability. Any associated mismanagement (such as overgrazing, combined with raindrop impact) may lead to high surface strength of soils [3, 7], impeding seedling establishment. Thus, grazing and use of fire are important management practices that can override the direct effects of clearing. A number of projects in northern Australia are or have been concerned with soil loss and runoff in grazed ecosystems [15, 23, 47] and these will be used in a project (LAMSAT) that will investigate the impact of land use on these processes in the semi-arid tropics [P. Hairsine, pers. comm.].

5.4. *Tree-grass relationships*

In grazing lands, trees are cleared in the expectation that pasture production will be increased. As pasture production is a determinant of stock carrying

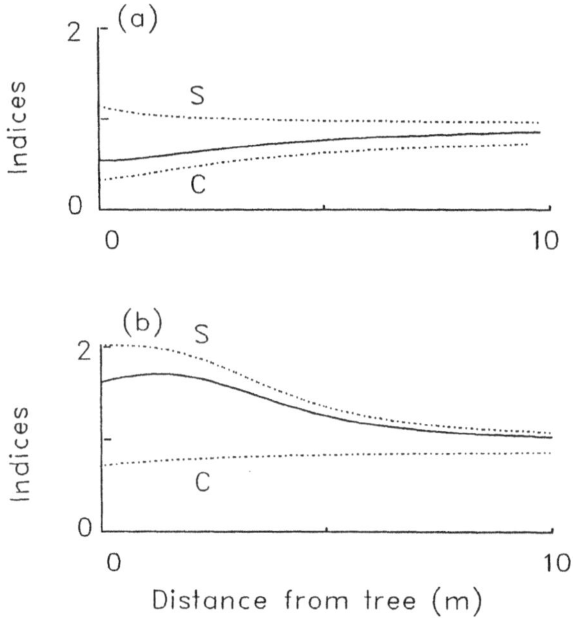

Fig. 5. The net effect of trees (solid line) on understorey pasture production, and the component stimulatory (S) and competitive (C) effects (adapted from [60]). In (a) trees have a net competitive effect on pasture growth, as is typical of many *Eucalyptus* spp. communities; and in (b) trees have a net stimulatory effect on understorey vegetation, as is often seen in nutrient poor situation with leguminous shrubs. These individual tree effects have different community level responses (see Fig. 6).

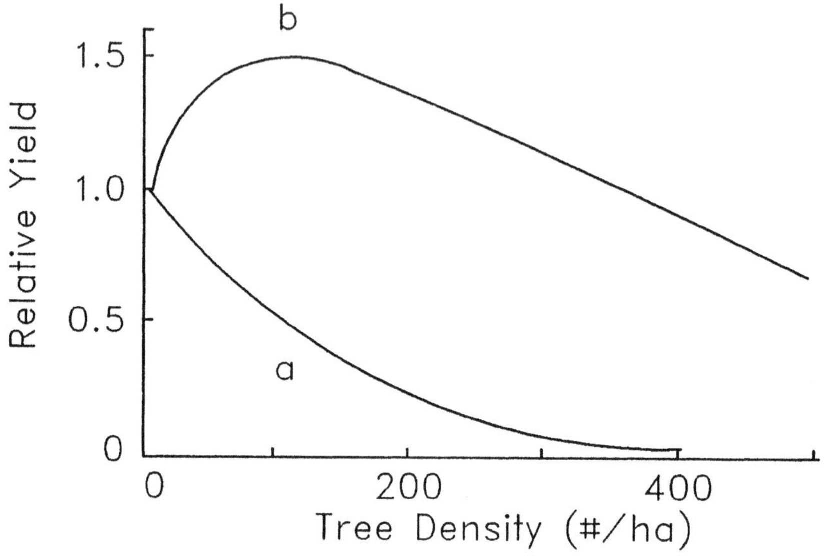

Fig. 6. The community level effects of tree density on pasture production for (a) situations where individual trees have a net competitive effect on the understorey; and (b) where the net individual tree effect is stimulation (adapted from [60]). These situations (a and b) are shown in Fig. 5. The reason for the decline in line (b) is that the stimulatory effect of trees reached some maximum whereas the competitive effects increase for each additional tree on the landscape. This results in some optimum value of tree density which also maximises pasture production.

capacity, a greater understanding of the impacts of trees on pasture is required. Recent simulation studies [60] have begun to develop this understanding for tropical and sub-tropical areas (Figs. 5 and 6). As seen in these figures, tree effects on pasture growth range from strong competition to stimulation. Studies on the effects of tree cover on understorey growth [76, 77] and a recent agroforestry study [12] are examples of the biological studies required to improve the understanding of tree-grass relationships.

Further studies are underway on the impact of rainfall distribution, soil fertility and soil water holding capacity on tree-grass balance. This work involves the inclusion of tree competition for water in a simulation model (GRASP) which estimates native pasture production given measurements of soil fertility and water holding capacity [38, 52]. Marked differences in the effect of trees on herbaceous understorey are due solely to differences in the amount and distribution of rainfall (Fig. 7). This wide range of responses makes the management of timbered areas for livestock production complex, as trees increase the variability in safe stocking rates of woodlands [62]. This is further complicated by the need to take into account the benefits of retaining trees for other purposes such as shelter and shade.

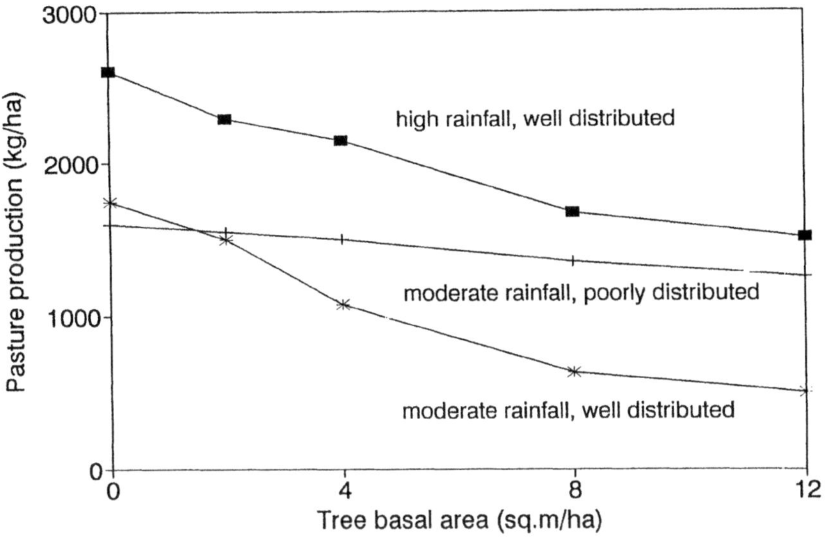

Fig. 7. Dry matter production of pasture at a range of tree basal areas for three selected years for a site near Duaringa, Qld. These results were simulated using a version of GRASP [38] modified to simulate the effects of trees [61]. All simulations used the same nutrient status and water holding capacities from a 12 year simulation period (1967—1978). Years were selected to show the range of responses observed.

6. Catchment level processes

Decisions that are appropriate at a property or individual paddock level may be totally in opposition to national goals. The clearing of a virgin *Acacia harpophylla*-softwood scrub in Queensland may be economically viable and may result in sustainable production, but it would be inappropriate as there is almost none of the vegetation type remaining in the state. The national level of organisation is not appropriate for making recommendations on what should be done on any particular management unit although it is necessary for setting general goals and targets for retention and for deciding national priorities for conservation. River catchments (or sub-catchments) are the most appropriate levels at which to plan and implement strategies for effective retention of native vegetation.

Tactics for individual land units can be devised to agree with catchment goals. Management decisions made at the catchment level ensure that regional hydrological balance or wildlife population survival goals can be met. Also, national goals (e.g. maintaining habitat diversity or ensuring suitable wildlife movement corridors) can be achieved by planning at this level. An increasing emphasis is being placed on the individual properties as all states move toward implementing property planning schemes. However, the most appropriate planning and implementation scale is the catchment, and insufficient attention has been placed on understanding processes at this level of organisation.

7. Future needs

7.1. *Legislation*

South Australia and Victoria have effective working systems regarding retention of native woody vegetation that other states may build upon and modify for their own needs. The current regulatory and incentive programs in these states provide a basic level of protection for all native vegetation. The challenges of the future in Victoria and South Australia are to provide native vegetation management programs to ensure the ongoing health of retained vegetation, and to make available financial and non-financial incentives to encourage improved management [39].

The range of agroclimatic zones present in Australia, and the vastly differing status of native vegetation among states will result in different legislative approaches being taken by the states. Some of the actions or approaches that should be considered in any legislation include:

- planning and implementation of overall controls should be at the catchment level with specific management and planning structures developed on a regional or sub-regional level (Recommendation # 3 [25]);
- all landholders should develop property *management* plans, i.e. these must be more than an inventory of natural resources and capital improvements (Recommendation # 4 [25]);
- all state departments and agencies involved in land and water use should have input into planning and implementation levels of any clearing proposals;
- local government should be enabled to provide rate rebates to landholders who are conserving areas of native vegetation;
- State Governments should be enabled to purchase, restructure and resell private land to conserve areas of remnant vegetation;
- tax relief should be provided for fencing and works designed to protect or manage native vegetation;
- counselling (financial and other) should be provided to rural landholders who are aggrieved by planning restrictions on native vegetation removal.

7.2. *Research*

Many issues concerning the management of native vegetation have to be resolved if our society is to develop a balance between conservation and agricultural production. These include the following:

Economics

- How can native vegetation management enhance property viability?
- How are appropriate values placed on intact native vegetation, and on the commercial products of forests and woodlands?
- How should an individual be compensated if added costs or lost income is associated with conservation?
- What are the most effective means (e.g. legislation, subsidies, penalty taxes) of encouraging conservation?

Fauna and flora

- What are the relationships between size, shape and degree of disturbance of vegetation blocks and long-term viability of flora and fauna?
- What fauna populations currently exist, what are desirable or acceptable populations and what are their specific management requirements?
- What is the value of native vegetation corridors and 'stepping stones' for movement of wildlife and how should these be maintained or established?
- What is the impact of exotic weeds, feral animals and agricultural chemicals on flora and fauna in remnant vegetation?
- We must improve our understanding of less-well researched vegetation communities and the fauna that depends on them, and must target programs at critical areas for conservation.

Management

- Can the potential problems with clearing-induced salting be predicted?
- How can reserved areas be managed to minimise plant and animal pest problems?
- How can native fauna (e.g. macropods) be managed to co-exist with agricultural production?
- What are the effects of fire management practices on remnant vegetation?
- Can degraded areas be rehabilitated?
- What are the effects of stock grazing on the regeneration of woody and herbaceous vegetation?
- We must improve the understanding of the management needs of native vegetation at all levels of government, statutory and semi-statutory agencies, and in the general community.

Monitoring

- Monitoring methods to assess the status of herbaceous and woody vegetation in remnant areas should be developed and the current condition of soils, pastures and woody vegetation on farms should be assessed.

Social

— We must improve the general understanding and acceptance of the importance of native vegetation amongst the community, especially in rural areas.

Legislative

— Various models of legislation need to be developed to provide for planning and control of native vegetation management at the catchment and regional levels.

8. Conclusion

Australia is at an important stage in its management of native vegetation resources. Cropping regions have little remnant vegetation in good condition and serious degradation is already obvious and further degradation is inevitable. Management in these areas must focus on maintaining what natural vegetation is present and rehabilitating degraded areas. Replanting areas with native woody vegetation is undoubtedly required. In grazing regions generally a high proportion of the original woody vegetation is intact. However, considerable damage has been done to herbaceous vegetation, and soil erosion is widespread. Introduced woody species are increasing in area and density, particularly in northern Australia, threatening both the natural heritage value and agricultural production.

Better integration of existing knowledge is required to develop land use plans that meet the multiple goals of agricultural production, hydrological balance and nature conservation. Economic aspects of native vegetation resource use must be studied to ensure that the goals above can be achieved.

Acknowledgements

Ann Prescott and Geoffrey Bishop of South Australia provided information contained within this paper for their state.

References

1. Abbott I (1983) The meaning of *z* in species-area regressions and the study of species turnover in island biogeography. Oikos 41: 385—390

2. Anon (1991) The Role of Trees in Sustainable Agriculture. A National Conference held at Albury NSW, October 1991. National Agroforestry Working Group
3. Arndt W (1965) The nature of the mechanical impedance of seedlings by surface seals. Aust J Soil Res 3: 45—54
4. Arnold GA (1990) Can kangaroos survive in the wheatbelt? West Aust J Agric 31: 14—17
5. Australian Survey and Land Information Group (1990) Atlas of Australian Resources Volume 6. Vegetation. Australian Government Publishing Service, Canberra
6. Bevege DI (1978) Biomass and nutrient distribution in indigenous forest ecosystems. Queensland Department of Forestry Technical Paper No. 6
7. Bridge BJ, Mott JJ and Hartigan RJ (1983) The formation of degraded areas in the dry savanna woodland of Northern Australia. Aust J Soil Res 21: 91—104
8. Burgess DMN (1989) Management burning of woody weeds: economics. Agfact P7.2.5 NSW Agriculture & Fisheries
9. Burrows WH (1991) Tree clearing in perspective — a question of balance. Search 22: 46—48
10. Burrows WH, Scanlan JC and Anderson ER (1988) Plant ecological relations in open forests, woodlands and shrublands. In: Burrows WH, Scanlan JC and Rutherford MT, eds, Native Pastures in Queensland: The Resources and Their Management. Queensland Department of Primary Industries, Brisbane, pp 72—90
11. Burrows WH, Carter JO, Scanlan JC and Anderson ER (1990) Management of savannas for livestock production in north-east Australia: contrasts across the tree-grass continuum. J Biogeog 17: 503—512
12. Cameron DM, Rance SJ, Jones RM, Charles-Edwards DA and Barnes A (1989) Project STAG: an experimental study in agroforestry. Aust J Agric Res 40: 699—714
13. Carter JO, Bolton MP and Cowan DC (1989) Prickly acacia: save dollars by early control measures. Queensl Agric J 115: 121—126
14. Chippendale JF (1990) An economic evaluation of rubber vine in Queensland: preliminary findings. Proceedings of the Inaugural Queensland Noxious Weeds Workshop 1990. Weed Society of Queensland, Brisbane
15. Ciesiolka C (1987) Catchment management in the Nogoa watershed. Aust Water Res Comm Research Project 80/128. Department of Resources and Energy
16. Dilshad M and Jonauskas PE (1990) Impact of land use on agricultural hydrology in the Northern Territory semi arid tropics. Workshop papers Erosion/productivity and Erosion Prediction. Proc 5th Australian Soil Conservation Conference, Perth
17. Dowling AJ, Webb AA and Scanlan JC (1986) Surface soil chemical and physical properties in a brigalow-Dawson gum forest, central Queensland. Aust J Ecol 11: 155—162
18. Ebersohn JP and Lucas P (1965) Trees and soil nutrients in south western Queensland. Queensl J Agric Anim Sci 22: 431—435
19. Eckersley R (1989) Regreening Australia: the environmental, economic and social benefits of reforestation. CSIRO Australia Occasional Paper No. 3
20. Ehrlich A (1990) Agricultural contributions to global warming. In: Leggett J, ed, Global Warming: The Greenpeace Report. Oxford University Press, Oxford, pp 400—420
21. Ferguson IS (1988) Review of Private Forestry in Tasmania. Private Forestry Council of Tasmania, Kings Meadows
22. Forest Industry Council (1990) Secure Futures for Forests and People. Forests and Forest Industry Council of Tasmania, Hobart
23. Gardener CJ, McIvor JG and Moneypenny R (1990) Dry tropical rangelands: solving one problem and creating another. Proc Ecol Soc Aust 16: 279—286
24. Gillard P, Williams J and Moneypenny R (1989) Tree clearing in the semi-arid tropics. Agric Sci 2: 34—39

25. Green R, Harris S and Throsby D (1991) Final Report — Agriculture. Ecologically Sustainable Development Working Groups. Australian Government Publishing Service, Canberra
26. Greenwood G and Gum E (1986) The State of Biological Resources in South Australia. Department of Environment and Planning, Adelaide
27. Harrington GN, Mills DMD, Pressland AJ and Hodgkinson KC (1984) Semi-arid woodlands In: Harrington GN, Wilson AD and Young MD, eds, Management of Australia's Rangelands, pp 189—207. CSIRO, Melbourne
28. Hobbs RJ (1987) Disturbance regimes in remnants of natural vegetation. In: Saunders DA, Arnold GW, Burbridge AA and Hopkins AJM, eds, Nature Conservation: The Role of Remnants of Native Vegetation. Surrey Beatty and Sons in association with CSIRO and CALM, Sydney, pp 233—240
29. Hobbs RJ (1991) Reintegrating fragmented landscapes — a proposed framework for the Western Australian wheatbelt. Retention and Replacement of Vegetation Workshop. Proc 5th Soil Conservation Conference, Perth (in press)
30. Hobbs RJ and Saunders DA (1991) Re-integrating fragmented landscapes — a preliminary framework for the Western Australian wheatbelt. J Envir Manage 33: 161—167
31. Howden SM, McKeon GM, Scanlan JC, Carter JO, White DH and Galbally IE (1991) Managing pastures in Northern Australia to minimise greenhouse gas emissions: adaptation of an existing simulation model. Proc Simulation Soc Aust, November 1991, pp. 168—178
32. Humphries SE, Groves RH and Mitchell DS (1991) Plant Invasions of Australian Ecosystems: A Status Review and Management Directions. Australian National Parks and Wildlife Service Endangered Species Project No 58
33. Inglis D and Davey L (1983) An Economic Comparison of Forestry and Dairying in Welcome Swamp. Tasmanian Forestry Commission, Hobart
34. Kirby MG and Blyth MJ (1987) Economic aspects of land degradation in Australia. Aust J Agric Econ 31: 154—174
35. Lang PJ and Kraehenbuehl DN (1987) Plants of Particular Conservation Significance in South Australia's Agricultural Regions: Interim Report. Department of Environment and Planning, Adelaide
36. Margules CR and Nicholls AO (1987) Assessing the conservation value of remnant habitat islands: Mallee patches on the western Eyre Peninsula, South Australia. In: Saunders DA, Arnold GW, Burbridge AA and Hopkins AJM, eds, Nature Conservation: The Role of Remnants of Native Vegetation. Surrey Beatty and Sons in association with CSIRO and CALM, Sydney, pp 89—102
37. Margules CR, Nicholls AO and Pressey RL (1988) Selecting networks of reserves to maximise biological diversity. Biol Conserv 43: 63—76
38. McKeon GM, Day KA, Howden SM, Mott JJ, Orr DM, Scattini WJ and Weston EJ (1990) Management for pastoral production in northern Australian savannas. J Biogeog 17: 355—372
39. McLellan R and Farago A (1991) Management of native vegetation on farms in Victoria Conference Proceedings, 'The Role of Trees in Sustainable Agriculture' Albury, October 1991
40. McMurray S (1983) An Investigation of Tree Decline on Tasmanian Farms. Master of Environmental Studies Thesis, University of Tasmania
41. Mentis MT (1991) Are multi-paddock grazing systems economically justifiable? J Grassl Soc South Afr 8: 29—34
42. Mitchell AA, McCarthy R and Hacker RB (1988) A Range Inventory and Condition Survey of Part of the W.A. Nullarbor Plain, 1974. Technical Bulletin 47, Department of Agriculture, South Perth
43. Mott JJ and Tothill JC (1984) Tropical and subtropical woodlands In: Harrington GN, Wilson AD and Young MD, eds, Management of Australia's Rangelands. CSIRO, Melbourne, pp 255—269

44. Myers N (1990) Tropical forests. In: Leggett J, ed, Global Warming: The Greenpeace Report. Oxford University Press, Oxford, pp 372—399
45. Orr S (1991) Managing Your Dry Forests. Private Forestry Council of Tasmania, Kings Meadows
46. Payne AL, Curry PJ and Spencer GF (1987) An Inventory and Condition Survey of Rangelands in the Carnarvon Basin. W.A. Technical Bulletin 73, Department of Agriculture, South Perth
47. Pressland AJ, Scanlan JC and Myles DJ (1991) Ground flora composition influences rainfall-runoff and soil loss in semi-arid tropical rangelands of Australia. Montpellier, Int Range Cong (in press)
48. Prinsley R (1992) Conference Overview. This issue
49. Private Forestry Council of Tasmania (1990) Private Forestry beyond 2000. Private Forestry Council of Tasmania, Kings Meadows
50. Richardson, Senator Graham (1988) Foreword. In: Saunders DA, Hopkins AJM and How RA, eds, Australian Ecosystems: 200 years of Degradation and Reconstruction. Proc Ecol Soc Aust, Vol 16: v—vi
51. Richley L and Cooper M (1989) Land Degradation and Soil Conservation in Tasmania. Tasmanian Department of Agriculture, Hobart
52. Rickert KG and McKeon GM (1982) Soil water balance model: WATSUP. Proc Aust Soc Anim Prod 14: 198—200
53. Rickson R, Saffigna P, Vanclay F and McTainsh G (1987) Social bases of farmers' responses to land degradation. In: Chisholm A and Dumsday R, eds, Land Degradation: Problems and Policies. Cambridge University Press, Cambridge, pp 187—200
54. Rose R and Cox A (1991) Australia's natural resources: optimising present and future use. Australian Bureau of Agricultural and Resource Economics Discussion Paper 91.5
55. Saunders DA (1989) Changes in the avifauna of a region, district and remnant as a result of fragmentation of native vegetation: the wheat belt of Western Australia — a case study. Biol Conserv 50: 99—135
56. Saunders DA and Curry PJ (1990) The impact of agricultural and pastoral industries on birds in the southern half of Western Australia: past, present and future. In: Saunders DA, Hopkins AJM and How RA, eds, Australian Ecosystems: 200 Years of Utilisation Degradation and Reconstruction. Proc Ecol Soc Aust, Vol 16: 303—321
57. Saunders DA and Ingram JA (1987) Factors affecting the survival of breeding populations of Carnaby's cockatoo *Calyptorhynchus funereus latirostris*. In: Saunders DA, Arnold GW, Burbridge AA and Hopkins AJM, eds, Nature Conservation: The Role of Remnants of Native Vegetation. Surrey, Beatty and Sons in association with CSIRO and CALM, Sydney, pp 249—258
58. Scanlan JC (1988) Managing tree and shrub populations. In: Burrows WH, Scanlan JC and Rutherford MT, eds, Native Pastures in Queensland: The Resources and Their Management. Queensland Department of Primary Industries, Brisbane, pp 91—111
59. Scanlan JC (1991) Woody overstorey and herbaceous understorey biomass in *Acacia harpophylla* (brigalow) woodlands. Aust J Ecol 16: 521—529
60. Scanlan JC (1992) A model of woody-herbaceous biomass relationships in eucalypt and mesquite communities. J Range Manage 45: 75—80
61. Scanlan JC and Burrows WH (1990) Woody overstorey impact on herbaceous understorey in *Eucalyptus* spp. communities in central Queensland. Aust J Ecol 15: 191—197
62. Scanlan JC, Wilson BJ and Anderson ER (1991) Sustaining productive pastures in the tropics 2. Managing woody vegetation in grazing lands. Trop Grassl 25: 85—90
63. Schofield NJ (1992) Salinity. This issue
64. Schofield NJ, Ruprecht JK and Loh IC (1988) The Impact of Agricultural Development on the Salinity of Surface Water Resources of South-west Western Australia. Report No. WS27, Water Authority of Western Australia, Perth
65. Tasmanian Forestry Commission (1982) Report on Survey of Forest Owner's Intentions 1982. Tasmanian Forestry Commission, Hobart

66. Thorburn PJ, Bowie BA and Lawrance PL (1991) Effects of land management on groundwater recharge determined from non-steady chloride profiles. J Hydrol 124: 43—58
67. Treespeak (1991) Direct Seedling Program. Treespeak No 33, March—April 1991
68. Usher MB (1987) Effects of fragmentation on communities and populations: a review with applications to wildlife conservation. In: Saunders DA, Arnold GW, Burbridge AA and Hopkins AJM, eds, Nature Conservation: The Role of Remnants of Native Vegetation. Surrey Beatty and Sons in association with CSIRO and CALM, Sydney, pp 103—121
69. Victorian Government (1991) Draft Decade of Land Care Plan, Victoria. Victorian State Government, Melbourne
70. Walker B and Weston EJ (1990) Pasture development in Queensland — a success story. Trop Grassl 24: 257—268
71. Wallace KJ and Moore SA (1987) Management of remnant bushland for nature conservation in agricultural areas of Southwestern Australia — operational and planning perspectives. In: Saunders DA, Arnold GW, Burbridge AA and Hopkins AJM, eds, Nature Conservation: The Role of Remnants of Native Vegetation. Surrey, Beatty and Sons Pty Ltd in association with CSIRO and CALM, Sydney, pp 259—268
72. Western Australian Parliament (1990a) Discussion Paper No. 1. South-west region of Western Australia. Select Committee into Land Conservation, Legislative Assembly, Perth
73. Western Australian Parliament (1990) Discussion Paper No. 2. Agricultural regions of Western Australia. Select Committee into Land Conservation, Legislative Assembly, Perth
74. Williams J and Chartres CJ (1991) Sustaining productive pastures in the tropics 1. Managing the soil resource. Trop Grassl 25: 73—84
75. Williams M and Goodwin D (1988) Conservation of biological diversity on the Fleurier Peninsula. South Australian Naturalist 63(2): 24—39 and 63(3): 64—66
76. Wilson JR, Catchpoole VR and Weier KL (1986) Stimulation of growth and nitrogen uptake by shading a rundown green panic pasture on brigalow clay soil. Trop Grassl 20: 134—143
77. Wilson JR, Hill K, Cameron DM and Shelton MM (1990) The growth of *Paspalum notatum* under the shade of a *Eucalyptus grandis* plantation canopy or in full sun. Trop Grassl 24: 24—38
78. Wolfe PM, Wright RG and Murphy DG (1990) Report of a Review of Land Policy and Administration in Queensland. Presented to Honourable AG Eaton, Minister for Land Management, Brisbane

Agroforestry Systems **20**: 167–186, 1992.

Integrating wood production into Australian farming systems

R. MOORE
Department of Conservation and Land Management, Busselton, Western Australia

Key words: agroforestry, sustainable agriculture, wood production, multiple purpose trees, tree crops, timberbelts, windbreaks

Abstract. About 21% of agricultural land in Australia is seriously degraded by salination and wind erosion. Trees can contribute to overcoming these problems. There is potential for trees to also produce wood, thereby diversifying farmers' incomes, reducing Australia's bill for importing wood and increasing exports of wood products. Even though many of these multiple purpose plantings would be widely scattered and growth rates low in some regions, Australian farmers should have a comparative advantage in wood production because trees also provide substantial landcare and agricultural productivity benefits. The establishment of new industries based on wood from farmland requires leadership from Government, planning at national, regional and local levels, and innovative techniques to finance planting. Removal of taxation disincentives for wood producers and the introduction of legislation to allow profit à prendre contracts would provide more favourable conditions for investment.

Introduction

A national study of land degradation undertaken in 1975–77 found that 51% of Australia's agricultural and pastoral land had been degraded to the extent that it required treatment [Woods, 1983]. It is widely accepted that trees can help to combat these problems [Eckersley, 1989]. Trees can help to prevent erosion of soil by wind by reducing wind speeds (Bird, this issue). Trees can also combat salinisation by lowering the level of ground-water (Schofield, this issue). The Institute of Foresters of Australia [1989] stated:

> Tree loss has been at the core of almost every aspect of land degradation in Australia. Tree replacement will be essential if we are to redevelop the fertility of many of our degraded soils and maintain sustainable systems of agriculture.

Australia therefore needs a modified farming system; a system in which trees are an integral part. Integrated systems not only prevent land degradation but they also produce more and can be more profitable. Bird [1988] concluded that 10% of land can be planted with shelterbelts without reducing agricultural productivity. Studies by Anderson et al. [1988] found that the combined production of widely spaced pine trees and grazing sheep was 30% greater than the production from sheep alone (convention agriculture without trees).

Agroforestry layouts can include belts, windbreaks, wide spaced trees and blocks. Strategies such as 'whole-farm planning' are available to integrate trees with techniques such as stubble retention, drainage and perennial pastures and to design layouts of trees which produce the maximum of benefits [Campbell, 1991a].

Australia currently imports about $2.2 billion of wood and wood products annually [ABARE, 1989]. Much of this is manufactured paper products but it indicates the potential for import substitution on the domestic market. The nation's substantial balance of trade deficit is also a strong reason for developing new export industries based on wood.

Some 10 to 25% of farmland requires trees in order to combat land degradation (Bird, this issue). On this basis 26 to 65 million hectares of the 260 million hectares of degraded land needs a restored tree cover. This area could produce a large wood resource, even though growth rates will generally be low. Furthermore some economic return from the planted trees will provide extra incentive for planting. This paper examines the potential for wood production, with the possibility of export markets, to be an integral part of Australian farming systems.

The potential for wood production from agroforestry

The overall scale of tree planting likely to be adopted by Australian farmers to gain landcare (i.e. measures to combat land degradation) benefits is very large. Such an area will produce a vast wood resource. However, the wood will be widely dispersed and transport costs will be a major factor in the viability of an industry based on wood products. Low value wood products will need to be processed locally. Hence much new farmland wood production will go into new industries requiring new local processing infra-structures and new market outlets.

Australian farmers should have a comparative advantage in wood production because trees will also provide substantial landcare and agricultural productivity benefits. It will be the aggregate benefit of growing trees that determines what is economical. This factor is particularly relevant when assessing the economics of producing wood in regions that receive less than 600 mm/yr rainfall. It follows that if the investment is to be funded from borrowings or by joint venture arrangements, the farmer may need to apply much of the wood revenue to service the investment and to take his profit in the form of the landcare and agricultural productivity benefits.

Types of wood products

The main wood products examined in this paper are reconstituted products such as particle board and medium density fibre-board, chipwood for pulp and paper, sawlogs, wood for energy (solid and liquid), and posts, rails and poles.

Reconstituted products

World consumption of reconstituted wood products such as particle board and medium density fibre-board is growing at a substantial rate [FAFPIC, 1990]. The increasing demand for reconstituted products is partly caused by a decline in the supply of sawlogs on the world market.

Particle board

Existing plants can meet the demand for particle board by the domestic market. Therefore new plants would need to target mainly export markets. The opportunities for developing new markets, particularly in Asia and the Pacific Rim, appear to be good [FAFPIC, 1990].

Calculations indicate that there may be potential to establish new particle board plants based on softwoods grown on farmland in areas that receive more than 450 mm/yr rainfall. Assuming processing plants require a log intake of 180 000 m^3/yr [NPAC Working Group 1, 1991] and that plantings (e.g. *P. pinaster* and *P. halepensis*) can produce 5 m^3/ha/yr [CALM, 1987 and R. Boardman, pers. comm.], the total area of softwoods per processing plant would need to be 36 000 ha. If 1.8% of the land was planted the required area of 36 000 ha could be obtained within 80 km of the processing plant. If 4.6% of the land was planted the area would be obtained within 50 km. There are many regions where such areas need to be committed to trees.

Medium density fibre-board

The case for new medium density fibre-board plants operating on a resource grown on farmland is similar to that for particle board. A suitable volume of logs can be grown within 80 km of a processing plant if rainfall is 450 to 600 mm/yr and just 1.8% of land is planted. In areas with less than 450 mm/yr rainfall, i.e. lower growth rates, a greater proportion of farmland would need to be planted to produce the required wood. Assuming that 5% of the landscape is planted with trees suitable for medium density fibre-board and that they produce 3 m^3/ha/yr, the required area of plantings could be obtained within 62 km of a central point.

Chips for pulp and paper

One of the most promising wood products for growing on farms in areas which receive more than 600 mm/yr rainfall is woodchips for pulp and paper. The demand on the international market for eucalypt chips for pulping is high. Figueiredo [1987] concluded that world consumption of paper and paperboard is increasing by 2.25% per annum. Chips from some eucalypts, particularly Eucalyptus globulus, are recognised as being superior for making high quality paper. An area of 200 000 to 300 000 ha of plantings would be required to fill this shortfall in supply to the Japanese industry. Thus there are strong prospects for establishing export industries based on eucalypts grown on farmland [FAFPIC, 1990].

In the south west of Western Australia about 10 000 hectares of *E. globulus* have been established on farmland in areas which receive more than 600 mm/yr. Many plantings are laid out in ways which fit with the overall needs of the farm; for example, to provide shelter or to combat salinisation [Bartle, 1991]. In Tasmania about 6000 hectares of eucalypts for chip have been planted [Prinsley, 1991b]. Similar projects are underway in the south east of South Australia, and south west and central Victoria. The relatively short term required to produce chipwood compared with other wood products is an advantage for farmers.

Sawlogs

There are good opportunities for Australian farmers to produce sawlogs. Two recent studies have examined the possibility of increasing Australia's wood production so that imports can be reduced, exports increased and the trade deficit reduced. The Australian Forest and Forest Products Industry Council [FAFPIC 1988] produced a plan which would result in a trade surplus in forest products by the year 2030. It called for the establishment of 523 000 ha of softwoods over 40 years and 76 000 ha of eucalypts over 20 years. The National Plantations Advisory Committee [NPAC) for the Federal Minister for Resources recommended that an additional 490 000 ha of softwood plantation would be required to replace imports of softwoods (1.67 million m^3/yr by the year 2030) [NPAC Working Group 1, 1991]. To replace imports of hardwood (360 000 m^3/yr), Australia would need to plant 112 500 ha of hardwoods [NPAC Working Group 1, 1991]. Clearly Australia needs to produce more sawlogs and farmers have the opportunity to contribute.

Softwood sawlogs

Australia has invested heavily in gaining knowledge and skills in the silviculture and utilisation of softwoods. Therefore it may be preferable, at least initially, for farmers to concentrate on the production of softwood sawlogs.

Modern softwood mills need an annual intake of at least 187 500 m^3 to be economic [NPAC Working Group 1, 1991]. Assuming a growth rate of 4 m^3/ha/yr (possible growth rate of softwoods in rainfall zone of about 500 mm/yr) 46 900 ha would be required. About 2.3% of farmland would need to be planted for a maximum transport distance of 80 km, and 5.9% for 50 km. In higher rainfall areas, where growth rates could be two or even three times greater, the total area of plantings required to produce the same volume of wood would be substantially less.

The main softwood species grown in south east and south west Australia is *Pinus radiata*. *P. pinaster* is grown on harsh sites (infertile sands or rainfall 450 to 700 mm/yr) in south west Western Australia. *P. caribaea* var. *hondurensis*, *P. elliottii* var. *elliottii* and the indigenous softwood *Araucaria cunninghamii* are grown in Queensland (sub-tropics). *P. halepensis* and *P. brutia* have potential in semi-arid areas (300 to 600 mm/yr) although breeding is required to improve stem form.

Hardwood sawlogs

The development of methods for milling small hardwood logs opens up opportunities for farm grown logs to be used. New milling and seasoning techniques had to be developed to deal with growth stresses in logs [Hillis and Brown, 1978]. Research carried out in Victoria and Tasmania in the Young Eucalypt Program, has developed techniques [Kerruish et al., 1991]. In Western Australia a laminated product called VALWOOD has been developed to use small logs [WURC, 1990]. The VALWOOD process uses thin boards laminated together in layers to make boards suitable for beams and furniture.

Some plantation grown hardwood sawlogs are already being used for high value products in Australia e.g. *E. grandis* (North Coast of New South Wales) and *E. regnans* (Central Gippsland in Victoria) [NPAC Working Group 1, 1991]. Other species with potential in high rainfall areas include *E. microcorys*, *E. pilularis*, *E. saligna*, and *E. maculata*.

There are opportunities for farmers to produce timbers for special purposes e.g. *Acacia melanoxylon* for high grade furniture timbers and *Santalum spicatum* (sandalwood) for incense manufacture. Some hardwoods indigenous to semi-arid zones (e.g. *E. loxophleba*) produce high quality solid wood, but more developmental work is required to identify markets, determine economics and evaluate other species. The general view is that markets for specialty timbers are limited but that on-farm processing will improve profitability [NPAC Working Group 1, 1991].

Posts and rails

The market for posts and rails is fairly limited. For a farm of 1000 ha with 50 km of fences and needing an annual replacement of 250 posts, an area of just 2 to 5 ha of trees could meet this demand. Thinnings from stands being managed for sawlogs, including stands of native forests, can be used for fence posts. If trees were planted primarily for other purposes such as controlling wind erosion, then it is likely that the costs of production would be less than the cost of buying posts.

The sapwood of most species needs to be preserved to make it durable because it comprises a large percentage of the cross-sectional area. There are simple and effective methods for preserving posts on farms as outlined in Cremer [1990] and CSIRO [1961].

Poles

The market for poles has traditionally been supplied from native hardwood forests on both private land and in state forest. The main users of poles are State electricity agencies who use them as power transmission poles. There may be opportunities for farmers to produce poles in plantations but the areas would need to be where tall dense forests can grow (rainfall >700 mm/yr). The main species being used for poles include *E. cloeziana* in Queensland and *E. marginata, E. muellerana*, and *Pinus radiata* in southern Australia. In general, poles are not likely to be an important wood product from farms.

Wood for energy

With the rising cost of energy and the growing concerns about the 'greenhouse effect' and the depletion of the ozone layer, the possibility of using wood for energy needs to be evaluated. Energy in wood can be used in two main forms; solid (fuelwood) and liquid (ethanol).

Fuelwood

The demand for fuelwood by the domestic market in the cooler parts of Australia has increased at a modest rate over the past two decades [NPAC Working Group 1, 1991]. The total consumption of domestic and industrial wood is 6.1 million tonnes annually. The retail price of fuelwood in urban areas exceeds $100 per tonne [Bulman, 1991], which is similar in cost to alternatives such as natural gas. Most fuelwood for domestic use comes from dead trees. As this resource is not sustainable, future supplies will have to come from residues of wood processing plants and from tree plantings.

There are opportunities for farmers close to centres of population to sell fuelwood. Dry fuelwood delivered to merchants fetches about $65 per tonne [Bulman, 1991]. Most farmers distant from markets are limited to producing fuelwood for their own needs.

Results of a study near Esperance on Western Australia's south coast (600 mm/yr) gave an indication of yield of fuelwood per tree and potential value (assuming a market exists). A 25-year-old two row *E. cladocalyx* windbreak produced more than 1.0 tonne of fuelwood per tree [D. Bicknell, pers. comm.]. At a nett value of $10 per tonne on the farm, this is $2000 per kilometer for a single row only. The second row was left to maintain shelter until the first row had regrown and the second row could then be harvested.

Solid fuelwood can also be converted into electricity using gas turbines [Hall et al., 1991]. It may be fashion that has prevented wood for energy from being considered seriously.

Ethanol

Woody biomass from trees and shrubs can be harvested and used to make ethanol by fermentation [Lynd et al., 1991]. Ethanol may be used neat or blended with conventional liquid fuels. In Brazil about 11 billion litres of ethanol are used annually. Expected improvements in fermentation technology should make ethanol cost-competitive with other fuels by the year 2000 [Lynd et al., 1991].

To be cost-competitive woody biomass should be transported only short distances. However ethanol can be economically produced in small local plants [Fox et al., 1990]. Plants could be established in towns to serve the local district. Ethanol produced could then be either used locally or exported. There is also scope to integrate the production of woody biomass with residues from other industries e.g. eucalyptus oil and tannins. This could improve the economy of both industries.

Wood for ethanol is therefore a potential wood product for farmers, including farmers in semi-arid areas. Belts of woody vegetation established primarily for shelter and control of wind erosion could be harvested and transported to an ethanol plant in a nearby town. The cut trees would sprout coppice from the stumps, thereby rejuvenating the belts with fresh foliage. Ethanol production from woody biomass warrants further investigation.

Management

Patterns of tree planting within farmland

Multiple purpose plantings of trees can be arranged in ways that fit with the overall requirements of the farm i.e. as part of a plan for the whole farm or even the whole catchment. The main ways of laying out trees are in plantations, timberbelts, windbreaks and stands of widely-spaced trees, as outlined in Cremer [1990] and Reid and Wilson [1985].

Managing trees for wood

The way trees are managed depends on the wood product to be produced, the species and site characteristics, although this may be modified by the other benefits expected from the trees. For example pruning trees in windbreaks may mean adjustments to standard forestry practices to ensure that the windbreak can still provide shelter [Moore, 1986].

The usual range of silvicultural techniques from establishment to harvesting are relevant to wood production on farms. Methods applicable to Australian conditions are discussed in Cremer [1990]. Practices for managing widely-spaced trees, a layout with special application in low rainfall zones, are outlined in Anderson et al. [1988].

Economic aspects

Farmers need to be able to compare the profitability of practices in which trees are integrated with farming with conventional agricultural practices without trees, and to determine what layouts of trees produce the greatest benefits. However, such assessments are difficult for several reasons, including the long lag between establishing trees and obtaining returns from wood, the multiple benefits of trees and the complex interactions between trees and agriculture.

Computer models can help take account of these many factors and are therefore a potentially useful aid in decision making. In Australia several agroforestry models have or are being built. The main ones are MULBUD [Etherington, 1983], TREE$PLAN [Bulman, 1991] and FARMTREE [Loane, 1991]. As the purpose of the models is similar, one model (FARMTREE) is examined to outline recent developments in this field.

FARMTREE has been developed in Victoria to help extension officers to evaluate the financial returns from agroforestry under different conditions. It allows the testing of different layouts, spacings, and pruning and harvesting regimes for a range of native and exotic trees. The model incorporates data on tree growth rates, response of trees to spacing, competitive effects of trees on pasture, and shelter effects on crops and livestock. Other data include cost of materials and log prices for different qualities and species.

Output from the model includes a number of variables e.g. log volumes, product types and values at each harvest, residual agricultural value in the competitive zone, crop improvement and lamb survival due to shelter. Internal Rates of Return and Net Present Values are calculated so that profitabilities of different combinations of trees and agriculture can be compared with each other and with conventional agriculture without trees.

A simulation of three different tree layouts of *Pinus radiata* (widely spaced, woodlot and timberbelt) on a sheep grazing enterprise in western Victoria in 700 mm/yr rainfall zone shows how FARMTREE can be used. The wide spaced pine (thinned to a final crop of 160 trees per hectare by 6 years of age), gave a slightly better rate of return than a conventional woodlot (4.8% compared with 4.2%) (see Table 1). The Net Present Values were

\$406/ha for the wide spaced stand and \$118/ha for the woodlot when discounted at 4% per annum. The benefits of larger tree diameters (and hence stumpages) and greater sheep carrying capacity with wide spacing outweighed the greater wood volumes and revenues (including revenue from thinnings) from the woodlot.

Table 1. Wood yields, net present values and internal rates of return for *P. radiata* grown in woodlot, timberbelt and wide spaced stand in Victoria. The values were produced by the FARMTREE computer model and are for a rainfall of 700 mm/yr, pulp price of \$5/m^3 and carrying capacity of 10 Dry Sheep Equivalents/ha.

Regime for *P. radiata*	MAI (m^3/ha/yr)	NPV (\$/ha)	IRR (%)
Woodlot	15.6	118	4.2
Timberbelt	15.7	7114	8.6
Wide spaced	8.3	406	4.8

The Mean Annual Increment (MAI) in wood growth of the wide spaced stand was 8.3 m^3/ha/yr compared with 15.6 m^3/ha/yr for the woodlot, owing to the lower degree of site occupancy and slower height growth. A three-row pine timberbelt produced a higher rate of return (8.6%) than both the wide spaced stand and the woodlot, despite having a much higher fencing cost. Improved sheep carrying capacity through shelter is one of the reasons for the higher rate of return.

The woodlot became more profitable (rate of return of 6.6%) than the wide spaced stand (6.3%) when several important parameters were changed; distance to pulp mill decreased from 120 to 30 km (and hence price for pulpwood increased), and rainfall increased from 700 to 900 mm/yr (increasing tree growth rate but sheep carrying capacity was maintained at 10 dry sheep equivalents/ha) (Table 2).

Australian species in woodlots or wide spaced layouts usually gave a lower

Table 2. Wood yields, net present values and internal rates of return for *P. radiata* grown in woodlot and wide spaced stand in two rainfall zones in Victoria. Values calculated by FARMTREE computer model.

Regime for *P. radiata*	MAI (m^3/ha/yr)	NPV (\$/ha)	IRR (%)
Woodlot — 700 mm/yr, pulp \$5/m3	15.6	118	4.2
— 900 mm/yr, pulp \$10/m3	20.0	1978	6.6
Wide spaced — 700 mm/yr, pulp \$5/m3	8.3	406	4.8
— 900 mm/yr, pulp \$10/m3	11.3	1537	6.3

rate of return than pine owing to lower growth rates and lower log prices in the 20 to 40 cm diameter range. An exception was *Eucalyptus regnans* in a woodlot in a high rainfall area (1200 mm/yr) producing pulpwood on a 25 year rotation which gave a 5.5% rate of return at a pulpwood price of $16/m^3$ (Table 3).

In lower rainfall areas (<550 mm/yr), timberbelts of native eucalypts such as *E. microcarpa* and *E. leucoxylon* gave the best returns. These species are suited to drier country and produce durable timbers which can either be sold or used on the farm. For example a three row belt of *Eucalyptus leucoxylon*, with other species for low shelter, produced a rate of return of 5.5% (Table 3). The timber returns came from thinnings at 10-year intervals for products of increasing value — firewood, posts, poles and sawlogs. The net impact on agricultural returns was positive because the shelter benefits outweighed the loss of grazing in the area fenced out for trees.

Table 3. Wood yields, net present values and internal rates of return for three wood growing options in Victoria; *P. radiata* woodlot, *Eucalyptus regnans* woodlot and *E. leucoxylon* timberbelt. Values calculated by the FARMTREE computer model.

Regime	MAI (m3/ha/yr)	NPV ($/ha)	IRR (%)
P. radiata — woodlot, 700 mm/yr	15.6	118	4.2
E. regnans — woodlot, 1200 mm/yr	24.9	680	5.5
E. leucoxylon — timberbelt, 550 mm/yr	6.5	3594	6.6

It must be stressed that the results are highly site specific i.e. returns and rankings of different regimes are affected greatly by site conditions. Furthermore, the data in the model are derived from a limited range of trials and there is uncertainty about many of the long term effects and market values. However, FARMTREE and the other computer models are important tools in the development of farming systems of which trees are an integral part.

Harvesting and processing wood

There is a range of options in the scale of harvesting and processing; from small farm based operations to large industrial systems. What scale of operations will be appropriate in the development of new industries based on wood production from farms?

On-farm processing

On-farm processing may be competitive and provide opportunity for diversification of the farm business. Farm equipment and labour can be used to fell, cut and stack logs for most products [Cremer, 1990]. Firewood for example can be cut, loaded and transported with the sort of equipment on most farms. Logs for chipping can also be felled, cut and stacked beside a road with minimum of special equipment. The farmer can then either transport the logs himself or pay somebody else to do it. All this can be scheduled for down time in other farm activities, hence reducing overheads. Farmers, if given the incentive, may also devise methods to reduce costs of transport, such as linking transport with backloads and transporting logs which have been sawn into large squared sections.

There is also potential for farmers to have logs milled on the farm using portable mills. Contractors with portable mills operate in some districts and more are likely to start operating as demand increases. On-farm milling avoids the expense of transporting whole logs and has the potential to increase profitability for farmers, especially those a long way from mills or with limited numbers of logs.

On-farm processing can also help overcome the psychological barrier to producing wood on farms. It enables farmers to learn and gain confidence about wood production.

In many cases farmers need to plan harvesting so that environmental and agricultural benefits of trees are maintained. Harvesting can be spread out over time to ensure that there is always a sufficient proportion of well developed trees on the farm. Tree arrangements and densities to provide non-wood benefits are outlined in other papers in this issue.

Industrial processing of wood

Large industrial processing may provide economies of scale. Softwood sawlogs, for example, can be processed efficiently in large modern mills. For nearby farmers it can be economic to sell sawlogs to sawmillers rather than milling them themselves. For other wood products, such as chip for pulp or boards, farmers may have no choice but to sell their small or low quality logs to large scale industrial plants.

Processing wood at a large scale plant can be more economical but there are several important constraints. They include acceptable transport distances, guaranteed and steady supply of wood, meshing of wood supply from Government and private growers and proximity to markets and ports. Failing to satisfy any one of these can make the venture unsuccessful for the farmer.

Encouraging new wood industries

The establishment of new wood industries requires conditions which favour investment. They include support from Government, reliable wood supply, integration with wood supply from native forest, and planning at national, regional and local levels.

The role of government

Government has an important role as a catalyst in establishing new industries based on wood from farms. For instance, Government can assist by providing 'seed' money for establishing a core resource and by facilitating the development of joint ventures with farmers. This helps in two main ways; firstly it shows industry that a resource of at least a certain size will be available, and secondly it provides farmers with some confidence that an industry will be established and therefore that it is worthwhile growing wood themselves. Joint ventures help mainly by reducing the high initial cost of establishing trees; a major disincentive for farmers. Governments can also assist by providing technical information and training.

Guaranteed and steady supply of wood

Wood processing companies will not invest in processing machinery unless they know that wood will be supplied at a steady rate over a lengthy period. Historically in Australia, State Forest Services and large private companies have been reliable suppliers of wood for industry. It is convenient for industry to buy wood from one supplier with a large wood resource. Future expansion into farmland will depend on developing methods to arrange supply from numerous farmers. Farmers and buyers therefore need to co-ordinate the growing and harvesting so that wood processing plants can receive a secure supply. Clearly there is a need for sound planning and legally secure arrangements to achieve this.

Combining with supply from native forest

In the past Australia obtained much of its wood from large mature trees in its indigenous forests. During the past two decades the situation has changed greatly. There is now little old growth forest left and what is left has mostly been made into conservation reserves. Thus logs harvested now are generally from small young trees in regrowth forests and are similar to logs grown on farmland. As mentioned earlier techniques for milling and seasoning small logs are being developed. Therefore, with planning, it should be possible to combine supplies from farmland and native forest.

Developing a national plan for producing wood

The National Plantations Advisory Committee recommended that a National Council be formed to develop a national plan for producing wood on farmland [NPAC, 1991]. The Council would oversee planning for new industries for processing wood, planning for combining wood supplies from farmland with that from native forests and developing new markets.

Developing regional plans for producing wood

Co-ordination at a regional level is essential for effective planning of new industries based on wood grown on farms. NPAC recommended the creation of regional committees to achieve this. Such committees have already been established in some regions; for instance the 'Albany 2002' Group in Western Australia. The task of regional committees is to collate and disseminate information on wood demand and supply, promote opportunities for new markets and investment, and identify needs for research and development. State agencies should provide support at the regional level.

Involving local farm groups

The success of new industries based on wood grown on farms depends on strong support from farmers. Thus, local farm groups are vital. Farm-tree groups and Landcare groups, of which there are over 900 across Australia [Campbell, 1991b], are well placed to assist. The National Plantations Advisory Committee recommended that '. . .local farm groups be encouraged, where appropriate, to broaden their role to consider wood supply and marketing. . . . .'. Input from these groups is essential if the needs of both local communities and wood processors are to be meshed. For example, there is a concern that local government authorities could perceive that the benefits of new wood industries might flow mainly to regional centres where wood processing facilities are located.

Incentives for farmers to produce wood

Assistance schemes

Most farmers need some sort of assistance to overcome the high cost of establishing and managing trees plus the burden of long wait for revenue. A range of financing mechanisms are being used in Australia, including joint ventures, afforestation investment companies and other contractual arrangements [Boutland et al., 1991].

Joint venture schemes

Joint venture or sharefarming schemes are agreements between landowners and investors. The agreements give the investor the right to the wood without having to purchase land. Landowners may be paid an annuity plus a share of returns at clearfelling, in any proportion. Schemes are operating in selected regions of Victoria, South Australia and Western Australia [Boutland et al., 1991; Prinsley, 1991b].

Examples of joint venture schemes are those offered by the Department of Conservation and Land Management (CALM) in Western Australia. Various schemes are on offer for softwoods and hardwoods. Under the softwood scheme the landowner enters into a Profit a Prendre contract with CALM. CALM plants and tends the trees and the landowner receives an annuity plus a 5% share of timber royalties at clearfelling. Annuities for softwoods range from $100 to $160 per hectare depending mainly on the amount of wood the site is likely to produce. Farmers can increase their annuity by contributing more, such as planting and tending the trees, or they can elect to take greater share of harvest revenue.

To facilitate multiple purpose plantings the Department of Conservation and Land Management has developed Timberbelt Sharefarming for eucalypts for chips. Under the scheme trees can be laid out in any configuration as long as belts are seven rows wide. Thus timberbelts can be laid out as an integral part of the farm to provide benefits such as shelter and control of salinisation. Timberbelts also use Profit a Prendre contracts but do not pay any annuity — the farmer takes harvest revenue in the same proportion as his share of costs. Prinsley [1991b] concluded that joint ventures without annuities are well suited to farmers who want to plant trees anyway for their benefits to farming e.g. shelter. Annuity schemes, particularly for large woodlots, appear useful 'where farmers are not interested in growing trees unless they can receive a direct annual financial return from them.'

Other joint venture schemes are available in Tasmania, Victoria, New South Wales and Western Australia with several of the major wood processing companies [Boutland et al., 1991]. They are mainly woodchipping companies.

Afforestation investment companies

Investors can take part in wood production through afforestation investment companies. These companies seek subscriptions from investors, usually by means of a prospectus. Substantial areas of plantations have been established in some regions in this way; for example, 20 000 ha of softwood plantations have been established near Tumut and Oberon in south east New South Wales.

Other contractual arrangements

Other contractual arrangements, such as grants and loans for tree planting, are offered by some state forest services and wood processing companies [Boutland et al., 1991]. The processors have first right to the wood.

The major shortcoming of most of the current assistance schemes is that they are for plantations only. Plantation type plantings are not an efficient way to gain landcare benefits and therefore have limited attraction to the majority of landowners. Schemes which assist plantings integrated with farming need to be developed. New schemes should also be established in lower rainfall areas (<600 mm/yr) where potential new wood industries have been identified.

Disincentives

There are a number of financial disincentives and structural impediments to wide-spread adoption of agroforestry practices. Major disincentives include taxation disincentives, pricing policies, export controls, legal constraints and planning constraints [ABARE, 1991]

Taxation disincentives
The method of taxing wood growing is a major financial disincentive because of the inequitable treatment of long term investments and the so-called period inequity. It has been shown that tax paid on investments in forestry is much greater than that paid on investments in other industries, including agriculture.

Period inequity arises in a progressive taxation system because a taxpayer with irregular income e.g. a wood producer, pays much more tax than one who has the same total income but which is more evenly spread over the same period. The National Plantations Advisory Committee made several recommendations to overcome these problems [ABARE, 1991].

Pricing policies

The system of pricing wood in Australia is a serious impediment. Investors lack confidence in the market price of wood. This is largely because of the domination of wood supply by State Forest Services. There are a multitude of Government restrictions on wood products from forests and rigid arrangements exist by which major forest industries buy wood, which tends to disadvantage the small grower. The National Plantations Advisory Committee recommended that State Governments allow market forces to operate in determining the price of wood.

Export controls

The Export Control Act 1982 is creating uncertainty about the right to export wood chips after the year 2000. The Act controls the export of wood chips from native forests. In particular the Act ensures that a fair market price is obtained, and that logs which could be processed into more valuable products (such as sawn timber) are not exported. The uncertainty arises because the Prime Minister announced in March 1991 that the export of wood chips would be phased out by the year 2000. It remains unclear whether this restriction applies to pulpwood produced on farmland.

Legal constraints

Legislation is needed to provide long term legal security to joint venture schemes. The New Zealand Forestry Rights Registration Act 1983 is a suitable model. This Act authorizes use of Profit à Prendre contracts and hence provides a 'forest right' on land with trees for wood without the need for costly survey and subdivision. Tasmania has the same legislation. It is important that other States adopt similar legislation, so that investors who do not want to buy land have easy legal security of wood on farmland.

Planning restrictions

Some State Governments and many local governments have tight controls over establishment of plantations because it seems to them to displace agriculture. In some areas pine plantation forestry is prohibited, and in others it incurs substantial penalties such as road maintenance costs and measures to control fire. Misunderstanding and ignorance may transfer such restrictions to integrated plantings. In order to prevent this, the National Plantations Advisory Committee recommended that government authorities be encouraged to make tree plantings for wood an 'as of right' use, where plantings are an integral part of the farm [NPAC, 1991].

Research and extension

For farmers in many parts of Australia who want wood production from tree plantings as well as other benefits, there is sufficient information for them to make a start. However there are substantial gaps in knowledge, particularly for the semi-arid regions [Prinsley, 1991a]. There is also a need to make current knowledge more readily available. The report of recommendations from the National Conference 'The Role of Trees in Sustainable Agriculture', outlines what steps need to be taken to overcome gaps in knowledge and barriers to action [Prinsley and Moore, 1992].

The main areas needing research include:

Tree breeding

- Select species with the greatest potential to produce useable wood and establish breeding programs to improve vigour and wood quality. There is enormous potential, through tree breeding, to improve particular wood properties so that milling and seasoning small hardwood logs are easier, and to increase tree growth rates.
- Test provenances of selected species on a range of farm sites on a regional basis across Australia. Build up data on which species and provenances are suited to what sites.

Optimising benefits of agroforestry

— Measure the total productivity of farming systems with trees.
— Determine hydrological, shelter and other non-wood values of trees integrated with farming.

Establishing trees

— Develop methods for controlling insect pests which threaten plantings e.g. wingless grasshoppers. Some pre-emptive work is also needed to identify species of trees resistant to insect attack.
— Document plantings by farmers and research workers (including failures) to build up records of experience.

Quantity and quality of wood

— Collate existing data and where necessary collect additional data on yield and quality of wood for selected species, including species in semi-arid regions for specialty timbers.
— Involve farmers in research on farms to resolve knowledge gaps and problems. Silvicultural systems research should be developed on farms in co-operation with farmers.

Economics of growing wood

— Continue to develop simulation models such as FARMTREE, for different regions, as an aid for extension workers. The models need to be able to integrate information on the major benefits and interactions including timber, shelter, salinity control and pasture responses.
— Carry out social and economic cost-benefit analyses of producing wood on farms.

Wood utilisation

— Improve methods of harvesting and transporting logs for farm use.
— Investigate national and regional opportunities for marketing wood.

Extension

— Collate and distribute available information to farmers and extension workers, especially information on selecting, establishing and managing trees for wood and marketing of wood grown on farms. Information needs to be presented with a regional focus.
— Promote joint-venture schemes with forestry companies or individuals whereby the initial costs are not such a burden to the farmer.
— Improve the training of extension officers.
— Encourage farmers to plant and manage trees for wood by making use of

demonstration sites and 'hands on' field days (e.g. thinning and pruning days for various agroforests).

— Provide funding for the further development of a national tree database, preferably TREEDAT which is being developed by the Commonwealth Scientific and Industrial Research Organization and the Queensland Department of Primary Industry. The database should include site conditions, climate, species and provenance details so that farmers and extension officers can make better decisions about matching species, site and product.

Conclusion

There are many opportunities for wood production from Australian farmland. The viability of wood crops is strongly linked to being multiple purpose i.e. integrated into agricultural systems. In regions which receive more than 600 mm/yr rainfall there is a range of wood products which farmers can grow. High quality sawlogs, particularly softwood sawlogs, are especially marketable in many districts and there are good prospects for establishing new export industries based on softwood thinnings for reconstituted wood products. There are opportunities to develop markets for hardwood sawlogs, using recently developed milling and seasoning techniques for young eucalypts. Pulp and paper industries based on hardwood chips from farmland are being vigorously developed in several regions in southern Australia. There are opportunities for this industry to expand.

Joint venture schemes between farmers and investors have been established in some regions and need to be expanded. They are an efficient means of financing plantings and they can help with orderly marketing. They are also well suited to irregular arrangement of trees on farmland.

There is scope for farmers to produce fuelwood, posts, rails and poles, but markets are limited. Growing these products for on-farm use is an important option.

Large scale opportunities are more limited in the 300 to 600 mm/yr rainfall zone. There is some potential to grow sawlogs of specialty timbers but markets have yet to be developed. Ethanol from woody biomass is a possible product and deserves more investigation.

The lack of established markets for farmers to sell wood grown on farms is a constraint. The National Plantations Advisory Committee considered that planning for new wood industries is required at three levels; national, regional and local. The Committee also considered that changes to taxation rules and pricing policies for wood would help create a more favourable environment for investment in wood.

Thus Australian farmers and national policy makers have the opportunity to make wood production a major force in improving both the sustainability and economics of agriculture.

Acknowledgements

I thank John Bartle, Roslyn Prinsley, Trevor Beetson, Pete Bulman, Peter Beatty and Graeme Siemon for their comments on drafts. I also thank Bill Loane for information on FARMTREE and John Kellas and Robin Yule for information from Victoria and Queensland, respectively.

References

ABARE (1989) Australian forest resources 1989. Report by Australian Bureau of Agricultural and Resource Economics. Aust Gov Publ Serv, Canberra

ABARE (1991) Financial mechanisms for and structural impediments to the development of commercial plantations. Report by Australian Bureau of Agricultural and Resource Economics, 48 pp

Anderson GW, Moore RW and Jenkins PJ (1988) The integration of pasture, livestock and widely-spaced pine in South West Western Australia. Agroforestry Systems 6: 195—211

Bartle J (1991) Tree crops for profit and land improvement. WA Jour of Agric 32: 11—17

Bird PR (1988) Financial gains of trees on farms through shelter. In: Proc International Forestry Conference for the Australian Bicentenary, Vol 2, Aust Forest Develop Institute, Albury-Wodonga, April 1988

Boutland A, Byron N and Prinsley R (1991) 1991 Directory of Assistance Schemes for Trees on Farms and Rural Vegetation. Greening Australia Ltd, Canberra

Bulman PA (1991a) Timber — an attractive crop for farmers in South Australia. In: Proceedings of National Conference — 'The Role of Trees in Sustainable Agriculture', Albury-Wodonga, October 1991

Bulman PA (1991b) TREE$PLAN — a model to compare farm tree designs in southern Australia. In: Agricultural System and Information Technology Newsletter, Bureau of Rural Resources 3, No 3, pp 16—18

CALM (1987) Pine trees for timber under WA conditions. Information Sheet No 2—87

Campbell A (1991a) Planning for Sustainable Farming. Lothian Books, Melbourne, 234 pp

Campbell A (1991b) Landcare: testing times. Second annual report of the National Landcare Facilitator to the National Soil Conservation Program. July 1991, 74 pp

Cremer KW (1990) Trees for Rural Australia. Inkata Press, Melbourne, 455 pp

CSIRO (1961) Round fence posts: their preservative treatment. CSIRO, Division of Forest Products, Leaflet No 12, 25 pp

Eckersley R (1989) Regreening Australia: the environmental, economic and social benefits of reforestation. CSIRO Occasional Paper No 3, 28 pp

Etherington DM and Matthews PJ (1983) Approaches to the economic evaluation of agroforestry farming systems. Agroforestry Systems 1: 347—360

FAFPIC (1988) The future direction of Australia's forestry and forest products industry. Forestry and Forest Products Industry Council, Melbourne, 12 pp

FAFPIC (1990) World demand for forest products — opportunities in the Asia Pacific region. Forestry and Forest Products Industry Council, Melbourne, 123 pp

Figueiredo RA (1987) Eucalyptus pulp: developments worldwide and supply potential. Proceedings Expo Paper Asia, Tokyo, June 1987

Fox DJ, Morjanoff PJ, Gray PP, Dunn NW and Marsden WL (1990) Process evaluation for the bioconversion of sugar cane bagasse to component sugars. Aust J Biotechnology 4(1): 22—25

Hall DO, Mynick HE and Williams RH (1991) Cooling the greenhouse with bioenergy. Nature 353: 11—12

Hillis WE and Brown AG (1978) Eucalypts for Wood Production. CSIRO, Canberra, 434 pp

Institute of Foresters of Australia (1989) Trees: their key role in rural land management. Submission to the House of Representatives Committee of Inquiry into Land Degradation in Australia, March 1989

Kerruish CM and Rawlins WHM (1991) The Young Eucalypt Report. CSIRO, Canberra, 272 pp

Kiddle G, Boardman R and van der Sommen F (1987) A study of growth and characteristics of woodlot and amenity tree planting in semi-arid rural South Australia. South Australian Woods and Forests Department, Unpublished report

Loane B (1991) Economic evaluation of farm trees — methodology and data for FARM-TREE model. Victorian Department of Agriculture and Department of Conservation and Environment, 55 pp

Lynd LR, Cushman JH, Nichols RJ and Wyman CE (1991) Fuel ethanol from cellulosic biomass. Science 251: 1318—23

Malajczuk G, Morrison D, Havel J, Anderson G and Moore R (1984). An economic study of agroforestry in the Manjimup region, Western Australia. Forests Department of Western Australia Tech Paper No 10

Moore RW (1986) Timber production from wind-breaks in the Esperance region. Trees and Natural Resources 28(4), pp. 18—20

NPAC Working Group 1 (1991) Integration of plantation development with regional utilisation of wood from native forests. In: Report of National Plantations Advisory Committee — 'Integrating forestry and farming'. Appendix A3

NPAC (1991) Integrating forestry and farming: commercial wood production on cleared agricultural land. Report of the National Plantations Advisory Committee, Department of Primary Industries and Energy, Canberra, 32 pp

Peck AJ, Thomas JF and Williamson DR (1983) Salinity issues: effects of man on salinity in Australia. Water 2000 Consultants Report No 8, Aust Gov Publ Serv, Canberra, 78 pp

Prinsley RT (1991a) Australian agroforestry — setting the scene for future research. Rural Industries Research and Development Corporation, Canberra, 90 pp

Prinsley RT (1991b) A review of private forestry assistance schemes on Australian farms. Bureau of Rural Resources, Canberra, 38 pp

Prinsley RT and Moore R (1992) Report of recommendations from the National Conference, 'The Role of Trees in Sustainable Agriculture'. Bureau of Rural Resources, Report R/1/92, Canberra, 105 pp

Reid R and Wilson G (1985) Agroforestry in Australia and New Zealand. Goddard and Dobson, Box Hill, 255 pp

Schofield NJ, Loh IC, Scott PR, Bartle JR, Ritson P, Bell RW, Borg H, Anson B and Moore R (1989) Vegetation strategies to reduce stream salinities of water resource catchments in south-west Western Australia. Water Authority of Western Australia, Report No WS 33, 81 pp

Wood Utilisation Research Centre (1990) Out of the woods. Concluding report on the small eucalypt processing study. Western Australian Department of Conservation and Land Management

FORESTRY SCIENCES

19. R.A. Leary: *Interaction Theory in Forest Ecology and Management.* 1985 ISBN 90-247-3220-4
20. S.P. Gessel (ed.): *Forest Site and Productivity.* 1986 ISBN 90-247-3284-0
21. T.C. Hennessey, P.M. Dougherty, S.V. Kossuth and J.D. Johnson (eds.): *Stress Physiology and Forest Productivity.* Proceedings of the Physiology Working Group, Technical Session, Society of American Foresters National Convention (Fort Collins, Colorado, USA, 1985). 1986 ISBN 90-247-3359-6
22. K.R. Shepherd: *Plantation Silviculture.* 1986 ISBN 90-247-3379-0
23. S. Sohlberg and V.E. Sokolov (eds.): *Practical Application of Remote Sensing in Forestry.* Proceedings of a Seminar on the Practical Application of Remote Sensing in Forestry (Jönköping, Sweden, 1985). 1986 ISBN 90-247-3392-8
24. J.M. Bonga and D.J. Durzan (eds.): *Cell and Tissue Culure in Forestry.* Volume 1: General Principles and Biotechnology. 1987 ISBN 90-247-3430-4
25. J.M. Bonga and D.J. Durzan (eds.): *Cell and Tissue Culure in Forestry.* Volume 2: Specific Principles and Methods: Growth and Development. 1987 ISBN 90-247-3431-2
26. J.M. Bonga and D.J. Durzan (eds.): *Cell and Tissue Culure in Forestry.* Volume 3: Case Histories: Gymnosperms, Angiosperms and Palms. 1987 ISBN 90-247-3432-0
    Set ISBN (Volumes 24-26) 90-247-3433-9
27. E.G. Richards (ed.): *Forestry and the Forest Industries: Past and Future.* Major Developments in the Forest and Forest Industries Sector Since 1947 in Europe, the USSR and North America. In Commemoration of the 40th Anniversary of the Timber Committee of the UNECE. 1987 ISBN 90-247-3592-0
28. S.V. Kossuth and S.D. Ross (eds.): *Hormonal Control of Tree Growth.* Proceedings of the Physiology Working Group, Technical Session, Society of American Foresters National Convention (Birmingham, Alabama, USA, 1986). 1987 ISBN 90-247-3621-8
29. U. Sundberg and C.R. Silversides: *Operational Efficiency in Forestry.*
    Vol. 1: Analysis. 1988 ISBN 90-247-3683-8
30. M.R. Ahuja (ed.): *Somatic Cell Genetics of Woody Plants.* Proceedings of the IUFRO Working Party S2.04-07 Somatic Cell Genetics (Grosshansdorf, Germany, 1987). 1988 . ISBN 90-247-3728-1
31. P.K.R. Nair (ed.): *Agroforestry Systems in the Tropics.* 1989 ISBN 90-247-3790-7
32. C.R. Silversides and U. Sundberg: *Operational Efficiency in Forestry.*
    Vol. 2: Practice. 1989 ISBN 0-7923-0063-7
    Set ISBN (Volumes 29 and 32) 90-247-3684-6
33. T.L. White and G.R. Hodge (eds.): *Predicting Breeding Values with Applications in Forest Tree Improvement.* 1989 ISBN 0-7923-0460-8
34. H.J. Welch: *The Conifer Manual.* Volume 1. 1991 ISBN 0-7923-0616-3
35. P.K.R. Nair, H.L. Gholz, M.L. Duryea (eds.): *Agroforestry Education and Training.* Present and Future. 1990 ISBN 0-7923-0864-6
36. M.L. Duryea and P.M. Dougherty (eds.): *Forest Regeneration Manual.* 1991 ISBN 0-7923-0960-X

FORESTRY SCIENCES

37. J.J.A. Janssen: *Mechanical Properties of Bamboo.* 1991 ISBN 0-7923-1260-0
38. J.M. Bonga and P. Von Aderkas: *In Vitro Culture of Trees.* 1992 ISBN 0-7923-1540-5
39. L. Fins, S.T. Friedman and J.V. Brotschol (eds.): *Handbook of Quantitative Forest Genetics.* 1992 ISBN 0-7923-1568-5
40. M.J. Kelty, B.C. Larson and C.D. Oliver (eds.): *The Ecology and Silviculture of Mixed-Species Forests.* A Festschrift for David M. Smith. 1992 ISBN 0-7923-1643-6
41. M.R. Ahuja (ed.): *Micropropagation of Woody Plants.* 1992 ISBN 0-7923-1807-2
42. W.T. Adams, S.H. Strauss, D.L. Copes and A.R. Griffin (eds.): *Population Genetics of Forest Trees.* Proceedings of an International Symposium (Corvallis, Oregon, USA, 1990). 1992 ISBN 0-7923-1857-9
43. R.T. Prinsley (ed.): *The Role of Trees in Sustainable Agriculture.* 1993 ISBN 0-7923-2030-1

KLUWER ACADEMIC PUBLISHERS – DORDRECHT / BOSTON / LONDON